AF824622

The Brothers Bellum

James Albright

What readers are saying about The Brothers Bellum

A skilled writer and expert pilot intimately familiar with operating large, crewed jet aircraft, James Albright draws from his deep experience as both a U.S. Air Force aviator and storyteller to deliver well-drawn characters involved in the massive bombing campaign credited with speeding an end to America's long and unpopular war in Vietnam.

For those curious about the in-flight responsibilities, challenges, concerns and dire confusions confronting B-52 flight crews in combat, this narrative puts you in one of the eight-engine bomber's ejection seats – most often that of the Electronic Warfare officer. It is his responsibility to provide the aircraft's primary defense against its greatest threat – surface-to-air missiles. Accomplishing that requires a mix of anticipation, close calculation, nerve and the isolation of critical signals smothered in a roar of electronic audio babble.

For those intrigued by the workings of a military command bureaucracy charged to subdue immediately a resilient, well-armed and ever more knowing enemy, it's here as well -- complete with overused, intractable tactics, self-protection of rank and pecking order, along with the dismissal of valuable input from warriors actually doing the fighting. A disheartening look inside a puzzle palace.

And helping bring focus to the personal toll of two uniformed brothers Bellum (Latin for "war") confronting the erred guidance directing the persistent but flawed campaign are the family members and sweethearts whose support steadily descends from encouragement to worry to devastation as nights of bombing take a terrible toll.

While most of the story focuses on the mechanics of modern war – know that the Stratofortress's systems and roles are continually upgraded to confront 21st Century adversaries -- it also provides reflection on the true purpose and essential failings of war itself. This is done through the questioning by a third brother Bellum, whose positions on the matter are tested and evolve through unanticipated suffering at the conflict's conclusion.

Although Linebacker II took place a half century ago, the questions, concerns, criticisms and relief arising in the narrative from that long-ago combat withdrawal are being repeated today as the U.S. military exits another controversial conflict. The Brothers Bellum reminds all that what has gone before often repeats, and that we need to take careful measure to avoid experiencing anew the errors and agonies of the past.

—Bill Garvey

Also by James Albright

International Flight Operations

Flight Lessons 1: Basic Flight

Flight Lessons 2: Advanced Flight

Flight Lessons 3: Experience

Flight Lessons 4: Leadership

Flight Lessons 5: People

ISBN: 978-1-7356475-3-1

The following story is a work of fiction based on actual historical events. The characters are not meant to reflect actual people, past or present. I have made every effort, however, to make the events correspond to what really happened. I believe the flying chronology is true to history, as is the bureaucratic nightmare created by the Strategic Air Command's operations staff. I have the highest respect for the crewmembers who flew during Operation Linebacker II and I dedicate this book to them.

Thailand, Laos, North and South Vietnam

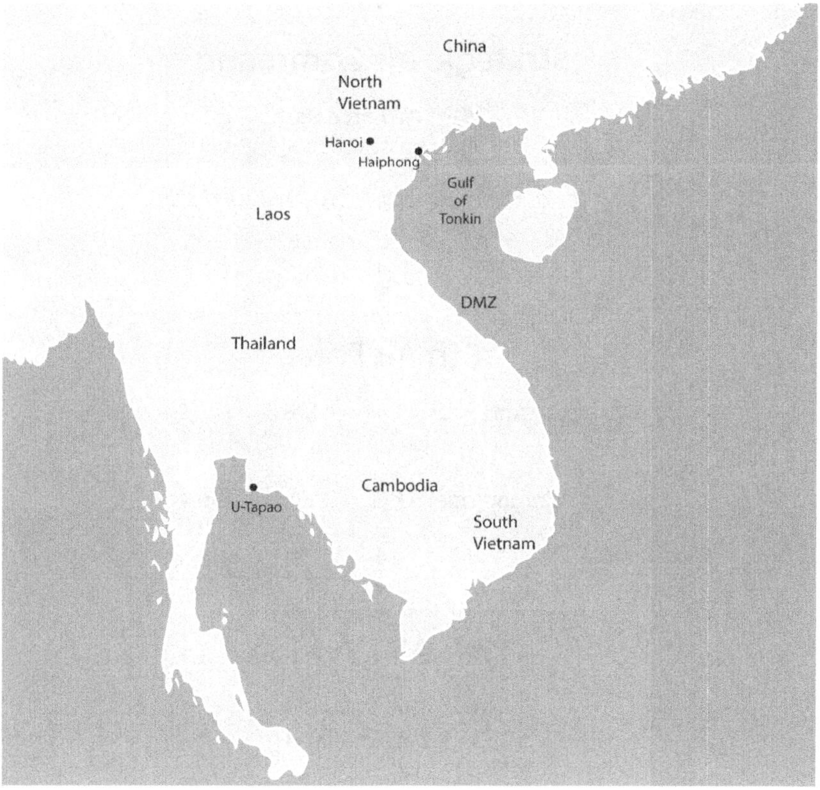

Credit: Author's drawing

The "as the crow flies" distance from U-Tapao to Hanoi is 580 nautical miles (nm). There are other important locations off the map. Andersen Air Force Base, Guam is 2,279 nm from Hanoi. Offutt Air Force Base, Omaha, Nebraska, is 6,884 nm from Hanoi.

The Chain of Command

The President of the United States

↓

Strategic Air Command

Offutt Air Force Base
Omaha, Nebraska

CINCSAC: Gen Avery Ham

↓

8th Air Force

Andersen Air Force Base
Guam

Commander: Lt Gen David Chesney

↓

17th Air Division

Royal Thai Navy Base
U-Tapao, Thailand

Commander: B Gen Kenneth Silvers

↓

307th Strategic Wing

Royal Thai Navy Base
U-Tapao, Thailand

Commander: Colonel Matthew Stipp

B-52 Cell Formation

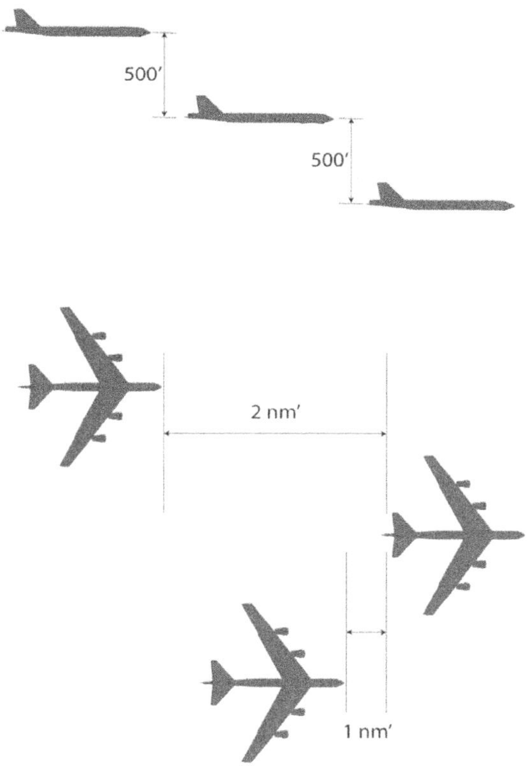

Credit: Author's drawing

The standard B-52 cell formation was flown with "number two" flying 1 nautical mile behind on a 45° angle to the right, and "number three" flying 2 nautical miles behind to the left. Position was maintained using radar but required extra maneuvering when turning, especially when the turn was opposite. Number 3, for example, was likely to fall behind in a turn to the right.

B-52D Cockpit Layout

23. EW OFFICER'S SEAT
24. EW OFFICER'S CONSOLE
25. CELESTIAL NAVIGATION SEAT
25A. PERISCOPIC SEXTANT CARRYING CASE
26. FILTERS FOR SCANNING LAMP
27. PERISCOPIC SEXTANT
28. SUNSHADES
29. SCANNING LAMP
30. CREW BUNK
31. ASSIST HANDLES
32. FOOD WARMING OVEN
33. INSTRUCTOR PILOT'S SEAT
34. FLIGHT REPORT HOLDER
35. CREW HOODS
36. PILOT'S SEAT
37. PILOT'S INSTRUMENT PANELS
38. WINDSHIELD WIPERS
39. PORTABLE OXYGEN BOTTLE
40. (Deleted)
41. (Deleted)
42. RADAR NAVIGATOR'S CONSOLE
43. RADAR NAVIGATOR'S SEAT
44. RELIEF TANK
45. COPILOT'S CLIP BOARD
46. COPILOT'S INSTRUMENT PANELS
47. COPILOT'S SEAT
48. PILOT'S OVERHEAD PANEL
49. NIGHT FLYING CURTAIN
50. BLACKOUT CURTAIN
51. SPARE LAMPS
52. DELETED
53. EW OFFICER'S SIDE TABLE (FOLDING)
54. CUP DISPENSER
55. DRINKING WATER CONTAINERS
56. CONTROL CABIN PRESSURE BULKHEAD DOOR
57. SHELF AND FOOD STOWAGE
58. INSTRUCTOR NAVIGATOR'S SAFETY BELT
59. CHEMICAL TOILET
60. INSTRUCTOR NAVIGATOR'S SEAT
61. LANDING GEAR LOCK STOWAGE
62. TRASH CONTAINER
63. HOT CUP
64. LADDER
65. LANDING GEAR LOCK STOWAGE
66. FORWARD ENTRY DOOR
67. NAVIGATOR'S SEAT
68. NAVIGATOR'S CONSOLE
69. STOWAGE HAMMOCKS
70. ALTERNATE LOCATION FOR SIGNAL LIGHT
71. SAFETY PIN AND STREAMER STOWAGE
72. THERMOS BOTTLE
73. TOILET PAPER HOLDER
74. SIGNAL LIGHT
75. GUNNER'S SEAT
76. GUNNER'S RELIEF TANK AND TUBE
77. AFT ENTRY DOOR
78. GUNNER'S COMPARTMENT PRESSURE BULKHEAD DOOR

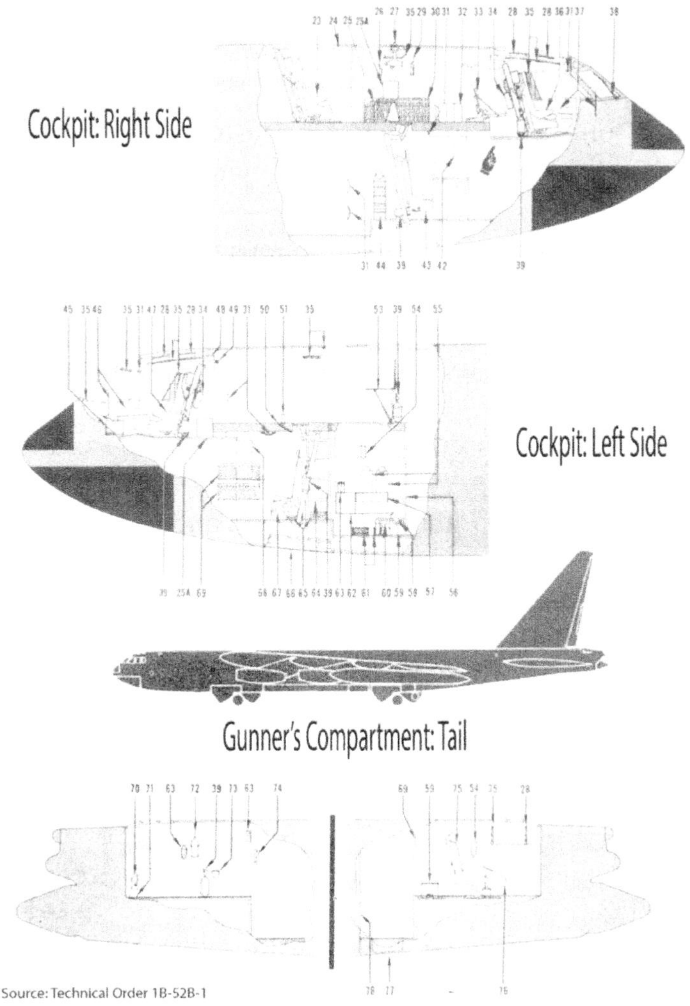

Credit: Technical Order 1B-52C-1, 10 July 1970

Crew positions in the B-52D. The layout was similar in the B-52G, except the gunner sat forward with the rest of the crew

Glossary

AAA – "Triple A" – Anti Aircraft Artillery

AN/ALE-20 – Flares

AN/ALE-24 – Chaff

AN/ALT-6B – Multi-Band jamming system used on about half the B-52Gs used during Linebacker II, manufactured by General Electric

AN/ALT-22 – An improved continuous wave jamming transmitter used on the B-52D and some B-52Gs, manufactured by General Electric, Burroughs, and Litton

AN/ALT-28 – Advanced jamming transmitter, manufactured by Hallicrafters

AN/ALR-20 – Radar detection set (No other information available)

AN/APR-25 – Six band radar detection and homing set, manufactured by Itek

Cell – The standard B-52 formation of three aircraft

Chaff – Small pieces of tinfoil-like metal strips used to jam enemy radar

Charlie – A highly qualified B-52 pilot positioned near the runway to make launch decisions

CINC – Commander in Chief

CINCSAC – Commander in Chief, Strategic Air Command

DMZ – Demilitarized Zone, the dividing line between North and South Vietnam

EB-66 – A USAF aircraft used specifically for jamming enemy radar

ECM – Electronic Counter Measures

EW – also EWO – Electronic Warfare Officer

Fan Song – SA-75's radar used to control the missile and guide it to its target

IP – Initial Point – Normally the start of the bomb run

KIA – Killed in Action

Linebacker I – The tactical bombing campaign from May to October 1972

Linebacker II – The bombing campaign from December 18 to December 29, 1972

MIA – Missing in Action

Mig – Mikoyan/Gurevich series of Soviet jet fighter aircraft

PDI – Pilot Direction Indicator, an instrument used by the B-52 navigator to let the pilot know where to fly during the bomb run

POW – Prisoner of War

PTT – Post Target Turn

ROE – Rules of Engagement

SA-2 Guideline – The NATO name given the SA-75

SA-75 Dvina – A SAM provided to the North Vietnamese by the Soviet Union

SAC – Strategic Air Command

SAM – Surface to Air Missile

SIGINT – Signals Intelligence

Spoon Rest – The SA-75's radar used to locate targets

Strobe – A single strip of light on a radarscope

TAC – Tactical Air Command

TOT – Time over Target

Uplink – Guidance signal transmitted from a ground radar site to a launched SAM

UT – U-Tapao

Weasel – also Wild Weasel – Aircraft used to attack SAM sights

1. Captain Ernest Bellum

U-Tapao, Thailand

It was her, he spotted her as she first stepped into the officer's club. She was wearing the standard United States Air Force olive drab fatigues, making her conspicuous in a sea of flight suits. It wasn't that women were unheard of at U-Tapao Royal Thai Navy Base, or even at any other U.S. base connected with the war in Vietnam, but she wasn't a nurse. Nurses were reserved for pilots; he didn't have a chance with them.

She had the classic new second lieutenant in a foreign environment look about her. She stood almost at attention, her hands neatly pressed to her sides while surveying the room with wide eyes and quick, darting glances. He knew that given another week, she would realize she was one of a few officer females in a sea of male officers far away from home. If he wanted a shot with her, he would have to act fast. He made his move, choosing a pace that was neither threatening in speed or attention grabbing because of a lack of it.

"Hey," he said, "I was in the weather briefing you gave this morning. New?"

"Yeah," she said, "does it show?"

"A little. How about I buy you your first combat beer?"

"That would be great." He gestured toward the table he'd already claimed as his own, his flight cap with captain's bars staking his territory and an already empty bottle of Singha. With a deft motion of his right hand, he pulled a chair for his guest and signaled for the nearest waiter for two more beers.

"Lesson one," he said, "you have a choice of Beer Singha or Beer Singha."

"That will be fine," she said. "How was the mission?"

"They don't say Singha Beer over here. It is Beer Singha. Or if you are really local, just Beer Sing. It has something to do with the dialect."

"The Red Cross gave us a book of phrases. It has a few hints about phraseology, but nothing about beer." As she spoke the waiter presented two frosty bottles and retrieved the empty. They clicked their bottles in salute and took their initial draughts. Her fatigues were neatly pressed, giving away the fact she was so new that she still cared about such things. His flight suit was wrinkle-free because he had been in it all day and the combination of sweat and humidity steamed any wrinkles out of it.

"My brother was stationed here," he said. "Stephen said they didn't start refrigerating the beer here on base until a couple of years ago. Savages."

"Your brother? Was he also a pilot?"

"It was scrubbed."

"What?"

"The mission. You asked about the mission. We got to the border when an electronic counter measures airplane over the target got lit up and the R.O.E. said abort. So, we aborted,"

"Oh," she said. She sat quietly, her eyes surveying the room, as if lost or looking for an exit. She looked about the right age for a second lieutenant, he thought. A college graduate would be at least 21 or 22, give another year for training would make her just under 25. But it was hard to say for certain. The Air Force rules against flashy makeup and for almost masculine hair styles tended to make all women in uniform look the same, but her youth was hard to disguise. She was probably a few years younger than he.

"I'm not a pilot," he finally said. "My brother is. But he isn't the smartest Bellum brother. But he isn't the dumbest either."

"I see," she said. "I should have recognized your wings. I remember thinking it odd that pilots have the vertical stripes at the center of their wings where you have the Air Force symbol. So, you are a navigator."

"You are a weather officer. Do they call you a 'W. O.' or maybe a 'wo' for short?"

"Meteorologist. I've never heard those other terms. But nothing would surprise me."

"It was a good briefing. Everything was as you said it would be. Except the clouds."

"The clouds? It was supposed to be solid undercast at 10,000 feet for your entire flight. Wasn't it?"

"E - W. Electronic Warfare officer."

"What?"

"That's what I do on the B-52. I operate all the electronic equipment that keeps us safe from enemy fighters, missiles, and triple-A. Most people call us 'EWs' or if you want to be lazy, 'e-dubs.' But I like the sound of E - W."

"Oh," she said, studying her beer bottle. It was still nearly full, but the condensation clearly showed where the line between beer and the hot, sticky air was. The officer's club was carved out of the Thai coastal jungle when the base was built, in the mid-1960's. Three of the four sides of the patio were open to the elements and provided a view of a mixture of fan palms and banana trees. The smell was tropical, but the noises from the nearby flightline dwarfed the sounds of the surf, less than a mile away.

"Beeps and squeaks. That's what some EWs say we do. We listen for beeps and squeaks on our receivers, and we answer with bings and pings." Her eyes continued the survey of the room, which didn't appear to offer any immediate rescue.

"If it was up to me," he continued, "we would have gone ahead and hit our target today. But it wasn't up to me. Mother SAC values these B-52s for the nuclear mission and they aren't making them anymore. The Rules of Engagement value the bird over the mission."

"What do your brothers do?"

"How do you know I have more than one?"

"You said your brother the pilot isn't the smartest or the dumbest. That implies there is another brother."

"Very good. Hardly anyone gets that." He smiled again, but this time the smile showed up in his eyes and that made her smile. As if on cue, a cool breeze wafted into the large patio where they were seated causing them to inhale the change in climate along with mood.

"Stephen is the oldest," he continued. "He was born before our dad got sent off to fly B-17s in Europe. Dad was a pilot too. Then comes me. Then comes Martin. Martin has the heart of a saint and the brain of a Saint Bernard."

She laughed and then blushed at the thought. She easily made it into the "good looking" category in Ernest's eyes, but most women did. Her hair was long enough that it had to be tied into a bun, since it wasn't allowed on her shoulders. She was smaller than average, some would say petite. And she had a pleasant, almost melodic voice. Ernest realized that the pace of his speech had slowed and that he must have sounded nervous before in comparison. He also sensed she had relaxed even as his voice's tempo slowed.

"Can I ask you something a little embarrassing?" she asked.

"It's pretty hard to embarrass an EW."

"Okay. Why do they call the B-52 a Buff?"

"You've seen the plane?" She nodded. "It is a Big, Ugly, Fat," he paused. "Fellow."

She laughed again and smiled warmly. Now her eyes were on his and the furtive glances away ceased.

"I was particularly interested in how you said 'I think' in your forecast. Most forecasters read you their forecasts with an air of certainty. You made it clear these forecasts were, well, forecasts."

"Meteorology isn't an exact science. I am just getting familiar with the area, but weather mainly depends on the air mass and the topography. We have imprecise measurements of the airmass over here, we have historical models for over there, and we do our best to predict the weather. There is variability and there are factors we don't even know about."

"That pretty much describes any science. Knowing there are things you don't know is the mark of true intelligence. Did they teach you that in Air Force meteorology school?"

"The University of Oklahoma, actually. I got my degree before I even considered the Air Force, but the job market, well you know." He nodded. "Officer Training School to the rescue! Ninety days later I'm a lieutenant and a year later here I am."

"Good news for us at U-T." He paused and then said, "I hear a pair of footsteps, characteristic of a tall, thin captain and a short, portly major. A copilot and a pilot." She looked up and saw them approach from behind Ernest, about 20 feet away. She started to articulate the obvious question, but it was too late.

"Hey, the weather girl!"

The two approaching pilots grabbed two chairs from nearby tables. "Don't mind if we do," one of the pilots said. He was the taller of the two, Captain Harry Broadmore, the crew's copilot. He wore his flight suit as if it were a set of pajamas. The pocket zippers were half zipped, the center zipper revealed a yellow tee shirt with the logo of a local bar, and his sleeves were rolled up to reveal the forearms and huge hands of an untamed gorilla. He positioned his chair back to the table and straddled it facing the table.

"Is Captain Bellum here regaling you with stories of our combat heroism today?" the other pilot asked. He was Major Walter Stevens, a shorter, grayer, and plumper version of Broadmore and the crew's aircraft commander.

"He has been quite charming," she said.

"Well, we have a crew meeting to conduct," Stevens said. "It won't take more than ten, fifteen minutes. But it is, as they say, 'hush, hush.'" As he spoke, two more officers in flight suits approached. "Here come the navigators."

"That's okay," she said. "I should be going. Thank you for the beer. Maybe we can do this again?"

"Sure," Ernest said. "We are flying tomorrow evening. How about this time the next evening?"

"See you then," she said. Ernest and his pilots watched as she left, walking a straight line to the door, one combat boot in front of the other. He felt that they were somehow stealing his view. Once the door closed behind her, the pilots faced Ernest.

"Thanks a lot guys," Ernest said. "I think maybe we had some kind of connection until you showed up."

"Sure you did," Stevens said. "What was her name, anyway?"

"Her name?" Ernest said. "I never asked."

"Typical!" Harry said. The other two officers pulled up two more chairs and all eyes turned to Stevens.

"Not much to report," he said. "The squadron had nothing to say about our abort, since the SAM call came from the EB-66s. You didn't see anything, Ernest?"

"The EB-66s were almost on top of the target," Ernest said. "Their gear is on a par with ours, but distance matters."

"But did you see anything?" Stevens repeated.

"The AN/ALR-20 displays all frequencies in the bands the North Vietnamese use," Ernest said. "A SAM should be reflected by two strobes side-by-side."

"Here we go," Captain Tommy Smallwood said. Tommy was the crew navigator and was about Ernest's age. Captain Kevin Long simply sat and smiled, as if preparing for the show. He was the radar navigator and was closer to Stevens in age than the rest of the crew.

"The AN/APR-25 Radar Warning Receiver will show a strobe within concentric rings," Ernest continued. "It can provide triple-A, SAM, or airborne interceptor warning."

"Did. You. See. Anything?" Major Stevens asked again.

"Of course, I saw 'anything.'" Ernest said. "But with three B-52s in a cell and at least one EB-66 all jamming, I didn't see the enemy radar specifically. I saw a lot of jamming. I heard a growl characteristic of the gunner's radar. I heard the low-frequency pulses of several long-distance search radars. So, I saw and heard a lot of things. But nothing that was directly threatening us at the time."

"All right," Stevens said. "That is what the other crews are saying. So, the squadron was okay with our abort and the fact our secondary target was right next to the primary was their fault, not ours. The only message from the wing is a slight change to the ROE: we will now be permitted limited maneuvering if fired upon."

"About time!" Harry said. "Time for some of that pilot magic, manno-e-sammo." He darted both hands about, miming an airplane outflying a missile.

"How can we hope to outmaneuver a SAM?" Tommy asked. "This is a Buff we are talking about."

"The missile has limited fuel," Stevens explained. "Even a small amount of maneuvering on our part requires a large maneuver on the missile's part. Once it runs out of fuel, it slows down rapidly, decreasing its ability to track us and hopefully allowing us to escape. Now the ROE says we can do that, we just need to make sure we get back on course in time to release the bombs."

Stevens scanned his pocket notebook. "Let's see here." He adjusted the glasses on the bridge of his nose. His age was betrayed not only by the gold oak leaves of his rank, but by his graying and disappearing hair.

A waiter stopped by their table and the four recently arrived officers ordered more Singha. Ernest declined and started to rise. "Another thing," Stevens said. Ernest shifted his weight back into his seat. "Squadron has a bone to pick with you on the flight logs."

Ernest sat quietly. The fact that he had to do them at all was an affront to him. The squadron procedure gave the duty to the EW since the other officers would be busy during the flight to and from the target. The EW had nothing to do, they said, other than just prior to, during, and just leaving the target.

"Every EW in recorded history before you logs crew times in a certain way," Stevens said. "The first line belongs to the aircraft commander, the second to the copilot. Then comes the radar navigator, the navigator, the EW, and the gunner. You always put my line on top, the gunner on the bottom, and jumble everyone else up at random. It is giving the typists who enter all the logs into computers fits." He paused, waiting for Ernest's explanation.

"So far this war, ten aircraft with EWs have been shot down," Ernest said.

"What?" Stevens asked.

"None of them B-52s," Harry said. Ernest ignored them both.

"Of those ten EWs," Ernest continued, "at least half are known to have survived. How many are actually in North Vietnamese POW camps today?"

"What?" Stevens said.

"How many?" Ernest repeated.

"How many?" Stevens asked.

"Zero," Ernest said.

Stevens appeared to be at his limit, but he was known for his long fuse. He took a swig of his Singha to help extinguish the fuse that had been lit. After it was clear he would be silent, Ernest continued. "Question 1. Why are there no EW POWs? Question 2. What does this have to do with me putting our names in random order on the aircraft forms?"

Stevens, perhaps sensing he had no recourse, simply shrugged his shoulders.

"Because the Soviets want to get their hands on any EWs to figure out how we do what we do," Ernest said. "We have reports of EWs being sent off to Russia where whatever knowledge they have is beaten out of them and then their brains are dissected to get whatever is left over." The crew sat silently. "The reason I put our names in random order, is if we are shot down and they somehow recover the form, they will be able to deduce that I am the EW from the form. I don't want to give them that. I have navigator wings just like Kevin and Tommy. If the three of us all claim to be navigators, they'll be less likely to deduce any of us are EWs."

"Makes sense," Kevin said.

"It does," echoed Tommy.

"Okay then," Major Stevens said. "Why don't you leave the crew section of the form blank until we are on the way home? After we leave bad guy territory."

"That will work," Ernest said. "Okay, I'll do that."

2. Major Stephen Bellum

Omaha, Nebraska

He felt almost self-conscious in the blue uniform with the shiny gold oak leaves of a major pinned to his collars. His nervousness was partly due to the fact he was approaching the big palace that was Headquarters Strategic Air Command. But mostly it was the fact he wasn't in a flight suit, and he was newly promoted. As a captain, he was one of many. Now as a major, he was in rarified air. The captains, some of whom were older, saluted him smartly. He felt almost an imposter, returning their salutes. His blue uniform still fit; he was happy to see upon his return from Southeast Asia. Too many of his peers came home to families aghast at their increased girth, thinking that perhaps combat duty suited them. The reality, all returning flyers knew, was the liberation of their waistlines in beltless flight suits.

The long walk from the massive parking lot made it clear he was about to embark on a new chapter in his life that had no prelude. Most of Offutt Air Force Base was hilly, owing to its position near the banks of the Missouri River. The headquarters building was placed on one of the highest points, making it tower over its subjects geographically as well as politically. Most of the building sat underground, but the three levels that rose above the earth filled the panorama. No doubt about it, Stephen thought, SAC was an Air Force of its own.

Once in the building, he came back down to earth. He realized that even as a major, he was again a small fish in very big pond. There were lieutenant colonels and colonels everywhere he turned. All in the same blue uniform.

Following the hand drawn map faxed to him at his temporary base quarters, he found his way to the top floor. With each floor the paint on the walls became brighter, the furniture became newer, and the personnel became more senior. On the top floor, the "low rent" to regal theme continued from the center stairwell to the outside offices with windows. The office of the Commander in Chief, Strategic Air Command, Action Group, occupied the penultimate position, next to CINCSAC's office. He opened the door and presented his orders to a sergeant at the front desk. "Thank you, sir," she said, picking up a phone. The office, destined to be his home for the next three years, was modern and spacious. The windows started about three feet from the floor and extended to the ceiling. They had a distant view of the flight line and the lone runway.

The sergeant cradled the phone and got up. "Please follow me, sir."

The eyes of nine other officers followed his path to the office nearest the windows and enclosed by its own walls: an office within an office.

The sergeant knocked on the open door. "Colonel," she said, "it's Major Bellum."

Stephen took two paces into the office, snapped his feet together and saluted. "Major Stephen Bellum, reporting for duty, sir."

The colonel returned his salute, smiled, and gestured to a seat opposite his own at the desk. "Have a seat, Stephen. 'Stephen' and not 'Steve,' right?"

"Yes, sir." The colonel wore the wings of a command pilot and appeared young for the rank. He was tall and trim and had a skin tone even lighter than his blond hair. Unlike the colonels back in U-Tapao, this colonel was neatly pressed from head to toe.

"I knew your dad before he retired," Colonel Grant Snedeker said. "Just two years ago, wasn't it?"

"Yes, sir."

"He worked in a different part of the building and I guess he was a generation before mine. He flew the B-17 in the big war."

"Out of England, sir."

"Good reputation. He held himself well. There aren't a lot of B-17 pilots left on active duty."

"He's kind of a legend in our family. There was no doubt the three of us sons would try to follow in his footsteps. But it didn't work out that way."

"Your younger brothers?"

"Ernest is an EW. In fact, he showed up at U-Tapao a week after I left. Martin, the baby, well he's kind of on his own path. He got a draft deferral and is a college student at Berkeley. But all three of us fully appreciate the fact our dad is a war hero."

"Yes, he is. I got in just after the war, so I missed the action. But I did the nuclear mission after that. You too, I see. G-models?"

"No sir. D-models all the way."

"Ah, I knew that, now that you remind me. I flew Gs on nuclear alert. It is a good airplane. How's she doing over 'Nam?"

"Not bad, I've heard. I think the D-model makes a better conventional bird, however."

"Well sure. Carries more bombs, of course. The so-called 'big belly' mod robbed the D of range, but she sure can carry a lot of iron."

"More than that, sir."

Snedeker raised an eyebrow and stared directly. "How so?"

"With the shorter tail, the G's nose tends to wander. That isn't a problem with a nuke, but dropping iron bombs, it throws off the accuracy. I hear if the radar navigator doesn't have the PDI exactly centered, and the pilot doesn't have it glued just right, the bomb is going to miss."

"PDI?"

"Pilot Direction Indicator, sir. Just a compass card controlled by the Radar Nav. He twists it to show the course, the pilot flies it to stay on course."

"Oh, PDI. Yes, I knew that. Now why is the G-model less accurate again?"

“They chopped the tail on the G, I think to reduce drag and give it more range.”

“Yes, of course.”

“A side effect of less tail is less lateral stability. Even with everything else steady, the nose on the G tends to gyrate in a little circle. It could be fractions of degrees. But from thirty-six thousand feet, those fractions add up.”

It took Snedeker a few seconds to process that. “I’ve never heard that.” But then his face brightened. “It’s things like this that will pay big dividends! CINCSAC said we needed someone with real ‘Nam experience. By God, he was right. Welcome aboard!”

“Thank you, sir, I’m looking forward to getting started.”

“Well get ready to drink from a fire hose, because we have a busy week for you. We have ten officers in the CINCSAC’s Action Group. Our job is to keep the four-star apprised of what is going on throughout the Strategic Air Command and carrying out his special projects. You a triple-burner?”

“Yes, sir.”

“Well that’s good and bad. Getting promoted three years early shows the system has faith in you. But you are the junior man on staff. Keep that in mind.”

“I will, sir.”

“You are the only officer on our staff who has actually dropped bombs in combat. So, if you think we are missing something, you speak up.”

“Absolutely, sir.”

“We’ve got five days this week for you to meet ten generals.” Stephen’s eyes lit up. “You’ll get used to it. Believe it or not, these generals will be more afraid of you than the other way around.”

“That is hard to believe, sir.”

“Well just think about it. You will have CINCSAC’s ear more than they will. Anything they say or fail to say can come back and bite them. It’s even worse these days. You know General Ham’s never flown a bomber?”

"Yes, sir. His exploits as a Korean War fighter pilot are the stuff of legend."

"Yup. President Nixon selected him because he thinks the Strategic Air Command is too timid to fight a conventional war. Our new CINCSAC has orders to fix that. Every other general officer in the building thinks their head is on the chopping block. In fact, you are going to meet your first one in fifteen minutes, so you better get moving. Remember, you represent CINCSAC, but you are just a major and you will accord each general with the appropriate level of courtesy. Clear?"

"Yes, sir." Snedeker handed Stephen a biography of his first appointment. "Brigadier General Howard Mayfield, Director of the Electronic Warfare Division."

"The highest-ranking EW in the command. He's not the most charismatic general you will ever meet. But he knows his stuff."

Stephen grabbed the bio and a map of the building. He had never met a general officer before, but the bio showed Mayfield was newly promoted with his first star. His office was in the first basement and didn't seem much bigger than Snedeker's. The man himself looked, in a word, mortal. He was a bit portly by Air Force officer standards, had thick glasses, and a face pocked with evidence of a lifetime's battle with acne. He rose from his desk and exposed a beer gut and a shirt button that had failed from the stress. He waved off Stephen's salute.

"No need for that," he said as Stephen reached for Mayfield's outstretched hand. "I have been looking forward to meeting you."

"Really?"

"Yes, yes. I knew your brother. I was commander of the Electronic Warfare Officer school at Keesler a few years ago. Ernest was our highest rated grad. Not just in his class. Ever. You must come from a family of geniuses."

"Just Ernest, I'm afraid. We knew he was on a level all his own before high school. He went to MIT on a full Air Force scholarship. I was surprised they put him in a cockpit."

"Too smart to be an E-dub?"

"Too smart for any flying job," Stephen said, realizing he may have insulted the general. "I thought they locked guys like him in the basement of the Pentagon. Solving the unsolvable. Things like that."

"Oh, I know what you mean. Did Ernest ever tell you about the final exam we give all EWs?"

"No, sir."

"It is intense. We put them on a scope with headphones. They get fifty tries to identify fifty different threats and they have to get them all. Takes most guys a few tries. Ernest breezed through it on his first try and in record time. So, we turned off the scope and made him do it again, by sound only. You are a Buff pilot, right?"

"Yes, sir."

"You've heard the term 'snake fangs' then?"

"A SAM."

"Yes. They call it that because on the scope, a SAM radar is seen as two high-frequency strobes side by side, growing from the bottom. It looks like the bottom fangs on a snake. But on audio, it sounds like a snake's rattler. There are nuances to each type of threat and some of those nuances can mean the difference between a threat detected, locked on, or defeated. Ernest got all fifty, on audio alone."

"I'm even more impressed with EWs than I was, hearing that," Stephen said, hoping to recover any lost ground in the general's eyes.

"Most pilots think the EWs sit back there sleeping until the bomb run."

"No that's not true."

"But it is! We EWs call it 'Hours and hours of boredom, interrupted by 15 minutes of sheer terror.' That's how the old crows describe their trips north of the De-Militarized Zone."

"That does bring up a question," Stephen interrupted. "Why are EWs called 'old crows' in the first place?"

"It goes back to the earliest practitioners in the art of electronic countermeasures, back in World War II. Hardly anyone understood radar and being able to counter it was next to magic.

Those operators were called 'Ravens.' In Greek mythology ravens were prophets. Over the years we became 'Crows,' and those of us who have seen action, 'Old Crows.' We became a society unto ourselves. The Air Force only added to our solitary reputation by making everything we do a level or two above Top Secret."

"That kind of rubbed me the wrong way the first time I commanded a Buff crew," Stephen admitted. "But once I understood just how complex it is, I was glad at least one member of the crew had a grasp on it."

"So far, no B-52 has ever been lost to a SAM. Back in '66 several were engaged, but our Electronic Counter Measures defeated them all. Back in April, fifty SAMs were launched against a B-52 attack and the results? All of them defeated. Only one Buff was damaged when the SAM that missed exploded nearby."

Which meant it didn't miss, Stephen thought, but did not say. "The ECM kept me safe while I was in theater, sir. Eighty-seven missions and not a scratch. I do have one question, however, general."

"Don't ever hesitate to ask me directly. You have full access to my staff and if they can't help you, you walk right into my office."

"Thank you, general. There is talk among the G-model crews that their jammers aren't as good as the D-model. Is that true?"

Stephen noticed that Mayfield appeared to be thinking, almost as if measuring his words before speaking. Stephen knew the issue was between two models of jamming transmitters. All D-models had the newer AN/ALT-22s, while half the G-model fleet was still stuck with the older AN/ALT-6Bs.

"To a pilot that may seem true," Mayfield said, pointing to the wings on Stephen's shirt. "The ALT-6Bs they are talking about have about half the power of the ALT-22s. To a pilot, more power is more better, right?"

"That's how we pilots think."

"Well, that's why we fly in cell formation. We find that three ALT-6Bs in a cell formation are just as good as the ALT-22."

"How do we know that for sure?"

"We've studied it."

After he left the EWD offices he realized the fallacy in the general's argument, but it was too late. "Stop being nervous," Stephen chastised himself. "Think!"

After two more meetings with two more generals, it was 1700 and it seemed the entire building left for the day. It was wartime on bankers' hours. With his schedule complete, he headed for his parent's home.

* * *

Stephen was for the moment what many called a "class two bachelor." His wife and kids were still at March Air Force Base, near Riverside, California, where he was stationed before U-Tapao. The assignment to Omaha came with just two weeks' notice and it was up to Susan to get the house ready for movers, the kids out of school, and the big drive east. They decided to wait until Christmas break, giving Stephen a chance to find a house on the market. It was all rather typical for an Air Force crewmember, but Stephen had an advantage. His parents lived in the nearby town of Bellevue, Nebraska, just a few minutes north of the base. He would be a frequent dinner guest until family life became normal again.

It was a middle-class home in a middle-class town. While the exterior appeared unique, a discerning eye would recognize that it was one of only twelve patterns repeated throughout the neighborhood. The four bedrooms remained much as they were when Stephen last lived there.

"It was pretty smart to rent in California," his mother said as they sat down for dinner. "Are you sure about buying here in Nebraska?"

"A little late for that question, Mary, he's already signed the contract," Jonathan said. "Besides, being at Headquarters SAC is good for his career, Stephen knows what he is doing."

"I hope so, dad," Stephen said. "Besides, buying a house here is a lot cheaper than anywhere in California."

"Did you see Ernest?" she asked.

"No," Stephen said. "But we talked a couple of times on the phone, and I have a friend keeping an eye on him."

"How is he?" she asked.

"Same old Ernest," he said. "He has everything figured out except how to act like a normal human being."

"Don't talk that way about him," she said.

"You know what I mean," he said.

"I do, actually," she said. "I wonder how you pilots can put up with his antics. He does seem to enjoy upsetting people."

"He's in the right job," Stephen said. "The EW is tucked away in his own cubby hole on the airplane. He doesn't have another pilot or navigator watching his every move. He isn't allowed to talk about what he does to anyone else, even his pilots."

"Not that he wants to talk to anyone," Jonathan said. "But I hear he is awfully good at his job."

"I hear that too," Stephen said.

"So, when are Susan and the kids coming?" Mary asked.

"The week before Christmas," Stephen said. "We want them to finish the term in school before having to drop into a new school here. Besides, it gives me a few months to concentrate on the job."

"It's a different kind of pressure in Building 500!" Jonathan said. "Isn't it?"

"Yes, it is," Stephen said. "I'm getting used to talking to general officers, but I am a little disappointed in some of them."

"You get used to it," Jonathan said. "A general is just a colonel who has the politics figured out."

"It is hard to keep from pointing out to a general that he doesn't make any sense," Stephen said. "Today I had a general who should know better tell me that three G-models with half-power jammers was as good as one D-model with a full power jammer."

"Apples to Oranges," Jonathan said. "You can't tell them that to their faces, of course. You got to get the message across so that they think they thought of it."

"Sounds like dealing with teenage boys," Mary said.

3. Martin Bellum

Berkeley, California

Even after three years on campus, he still felt like an outsider. Being on draft deferment did that to some men his age. They either became vocal opponents of the war or they tried to sink into the woodwork, hoping to escape notice. Martin Bellum had been in the latter category, but he knew his time was coming. His hair was longer than that of his military brothers but not so long as to call attention to his age. He favored blue jeans, but he kept them neat. You could have mistaken him for just another college student from a decade earlier.

He scanned the nearly empty meeting hall. There were small tables scattered about, as if at random. Not all the tables had chairs and there were chairs stacked against the wall. It was, in a word, messy. The campus newspaper had a seemingly permanent notice about an anti-war group's weekly meeting, but there didn't appear to be a meeting in progress at the promised time or location. There was an unattended beer keg and a few empty tables, so he poured himself a drink and sat at a corner table. He had barely started on the beer when she approached.

"New here?" she asked.

"Yeah." She smiled and sat.

"I'm Sherrie."

"Martin," he said, smiling.

She looked younger by a few years. Her brown hair was in need of a brush, her jeans in need of a wash, but her loose-fitting and bra-less blouse was perfect. Martin tried very hard to keep his eyes on her eyes or his beer, anywhere but the blouse. She pulled out the day's issue of the campus paper.

"Are you here because of this?" she asked, pointing to the headline on the paper, announcing the latest draft board results.

"Kind of."

"Let's see, what's your birthday?"

"February 13, 1951. But it doesn't . . ."

"I got it right here," she said, interrupting. She read down a list. "Good news. Your draft number is 118 and they are only calling up the first 95." She smiled brightly, perhaps thinking this was the news he wanted desperately to hear.

"Those are the draft numbers for men born in 1953. For my year, my draft number was 13 and they called up everyone with the number 125 or lower. I got drafted two years ago."

"Oh."

"My deferment ends when I graduate next June. Unless Nixon follows through with his promise, I am going to be a common Army soldier next year." He emphasized the word "common."

"That is awful."

"It seems I don't have a choice in the matter."

"I think you have options. You need to talk to Gene. I'll find him for you."

Martin sat back in his chair and noticed the room had suddenly filled with more men, his age and younger. Most were clutching the same issue of the campus newspaper, many looked anxious, but most had the same look of resignation that he had felt for the last two years. He felt somehow diminished by the apparent bond with his fellow draftees. It was becoming his role as a college student, just as that of many men his age now playing the role of army soldier.

Sherrie returned, followed by a man of medium built and height. His long hair was typical for Berkeley, but more neatly combed than what would be considered campus chic. He was carrying a hand drawn sign attached to a stake. "One point peace plan: Get out now!" He put the sign down next to the table. She retook her seat and he pulled up another and offered his hand. "Gene Phillips," he said.

"Martin Bellum. Are you going to tell me about Canada?"

"That used to be the best option, but not anymore, there are easier ways. If you go to Canada, you can never come back. But now if you don't want to be a baby killer, you can declare yourself a conscientious objector and volunteer for some other public service. It used to take a letter from the Pope himself to become a CO, but they made it easier." Gene seemed confident, self-assured, and uncommon. Martin studied the man's eyes, suddenly realizing that Gene was doing the same thing to him.

"What kind of public service?"

"You can join the Peace Corps. Or there are lots of other things. My brother got CO status last year and he works in a hospital in Philadelphia, cleaning bed pans and stuff like that. He has to do that for two and a half years, then he gets his life back."

"It can't be that easy," Sherrie said.

"Oh, it isn't easy," Gene said. "You have to write an essay and then you spend an hour in front of a panel of war vets who have to be convinced that, A, you have a conscience, and B, you are sincere. But my brother got away with it and he has no conscience. But he put on an act of sincerity that could have won an Academy Award."

"I think it would kill my mom and dad to have a CO son," Martin said. "Everyone thinks my dad was a war hero, and my two older brothers are following in his footsteps."

"You can use that in your essay," Gene said. "Youngest child traumatized by war monger family; that kind of thing."

"I don't know," Martin said. "Everything happens for a reason. I got drafted. I got deferred. Maybe I'll end up in Canada or as a CO. Who knows? Maybe McGovern can pull this off. He'll stop the draft."

"McGovern was a bomber pilot during the Second World War," Gene said. "He has blood on his hands. You can't trust any of them. Besides, it's immoral to support the war and you have to resist whenever you can."

"I know that. I'm a history major. I study these things."

"So, you should know."

"I do know. But I don't want any stigma attached to my brothers."

"They are as guilty as anyone else by wearing the uniform."

"No that's not true. They are just little pawns in the big man's game. You can't blame them."

Gene started to speak but stopped. The two men sat and studied their beer bottles, the marks on the table left by past beer bottles over the years, and just about everything within sight except each other. Sherrie shifted in her seat nervously, expecting more but realizing an impasse had been reached.

"Martin, are you close to your brothers?" Sherrie asked finally, breaking the silence.

"Not really," Martin said. "Stephen is practically an old man. He's almost thirty, has a wife, three kids, and a dog. Ernest is younger, but still four years ahead of me. He's a good guy, but a little weird. I think I was his only friend growing up. They are both in the Air Force. How would it look to have a military father with two military sons and a third son dodging the draft?"

"Well, I submitted my essay four months ago and they just now got around to scheduling me for my interview," Gene said. "This is, what, October? So, we have eight months until we graduate. If you are going to be a conscientious objector, you need to submit your essay sooner than later."

"I'll wait for the election," Martin said. "I don't want to put my family through hell if I don't have to."

4. Aborted

U-Tapao, Thailand

"I'm gonna get me a Mig today," Staff Sergeant John Polanshek said, rubbing his hands together. He was the gunner, the only enlisted man on the crew, and almost as young as Ernest. "No bomber has a fighter kill since Korea. Today we are gonna change that!"

"I just want to bomb something," Ernest said.

"Amen to that," said Tommy.

"You two rookies will bomb something soon enough," Major Stevens said. "The Rules of Engagement are there for a reason."

The crew bus pulled up to their bomber and each member of the crew completed the choreography that is a normal B-52 preflight. Ernest stowed his metal box of secrets below his desk in the "cave" that was his part of the cockpit, seated about 15 feet behind but on the same level as the pilots. It was something of an irony that he and the gunner were the only crewmembers dedicated to aircraft defense and were the two crewmembers seated furthest aft. Of course, the gunner sat a hundred feet further aft in the tail of the aircraft.

The smell of the airplane no longer caused him to have to restrain a guttural heave. Aircraft 55-0679 was nearly twenty years old and the years of bomber crews sweating in ejection seats left an "aroma" any B-52 crewmember would recognize. Boeing provided heating and cooling for the crew, but the EW compartment was a cockpit all its own and the ventilation was wanting. If the cockpit was cold, the EW compartment was hot, and vice versa.

Ernest lowered himself through the aircraft's lower hatch and walked the length of the aircraft, checking each of the many antennas dedicated to his ECM gear along the underside of the fuselage. The aircraft hadn't aged well and the skin along the sides and underneath was wrinkled from the stresses of flight. It was somewhat of an irony that the ECM antennas were the shiniest pieces of equipment on the airplane. From his view, kneeling on the tarmac, he could make out each antenna forward of the bomb bay. His view aft of that was obscured by Kevin Long inspecting the safeties on each of his bombs. Behind that he caught a view of Polanshek inspecting the belts feeding his four fifty-caliber machine guns.

Passing under the left wing, Ernest noted the conditions of each of four AN/ALE-24 chaff dispensers. After each, he said, "what a waste," in a personal ritual of his. He repeated the statement for each of the four other chaff dispensers on the right side of the aircraft.

Reboarding the aircraft, he picked up his helmet bag right where he left it and climbed past the "Black Hole" which was the navigators' station, seeing both Kevin and Tommy spreading charts across their shared desk. Tommy was technically the navigator, but Kevin would keep an eye on him. Kevin's title was "radar navigator," but really, he was the bombardier. Unlike the bombers of Ernest's father's era, these bombardiers didn't need large glass windows to aim their bombs. In fact, they had no windows at all.

As soon as the power cart was hooked up and the aircraft's electronics came to life, Ernest ensured all his gear at least passed the initial tests: the lights came on, anything with a test button passed the test, anything without wasn't obviously broken. He wasn't allowed to fire up any of his jamming gear because it would pretty much put the base out of business. "I'll never make that mistake . . ." long pause ". . . again," he said in another ritual designed to keep these kinds of things at the forefront of his mind. It was a risky business, dropping bombs. But he *wanted* to go.

"Copilot, EW," he said over the interphone, "I'm down one ALT-32L, the other two are operational. EW is a go."

He then leaned into the scope of his AN/ALR-20A search receiver, closed his eyes, and imagined the sounds to come. He reached to each of eleven jammer control heads and placed two fingers on each knob. Reaching to his left, skipping the broken jammer, he settled on the second AN/ALT-32L low-band jammer, rotating the second knob on the right, imagining how the scope would react. He repeated the process with all of his equipment, even the chaff dispenser control heads he wasn't allowed to use. His was a dimly lit world, only the backlighting on each instrument and the green glare of the two scopes in front of him would illuminate anything at all. He knew every instrument panel by feel alone.

"Engine start," Major Stevens announced to the crew. "Here we go, starting number four."

Ernest sat back in his seat, closed his eyes, and counted engines. With just one engine turning, the harmonics were all wrong and the fuselage shuddered in complaint. Once that engine was started, the order seemed almost random and seemed to vary with aircraft. With an unequal number of engines turning on one side of the airplane, the pulse of the vibration would rise and fall every two seconds. The pilots called it "wow-wow," but Ernest knew the phenomenon as reverb. But once the eighth engine came up to speed, it was as if the screeching was replaced by a barbershop quartet. Or octet.

It was an exercise in sounds for Ernest. The wheel brakes gave a particular complaint when the aircraft was stopped or even turned too quickly. The hydraulic system dutifully squealed when providing the muscles to move the flaps. Even the electrical system could be monitored by the pulsing of the fan over Ernest's head. His compartment was always too hot until they were airborne. But that part never came soon enough.

"Lead is off," Stevens said. Even from hundreds of feet away, the noise and vibration of the lead aircraft's takeoff thrust shook the ground and eclipsed the noise of their own aircraft. "There goes two." Ninety seconds later, he could hear and feel the pitch of the engines of his aircraft lower as the water injection system started pumping hundreds of gallons into the fuel and air mixture of all eight engines. The sound was deafening and even the copilot's speed callouts on the interphone became next to impossible to hear. One hundred and ten seconds later, when the engines ran out of water, it was as if they were shut down completely, it was so quiet by comparison.

Ernest shifted his eyes to his largest scope, looking for the radar signature of the two other aircraft in their cell formation. Their gunners' radars should be active and easily tracked. The signal was relatively faint, but Ernest knew where to look. As the third aircraft in a cell formation of three, he knew lead should be two miles ahead and to the right and number two should be off lead's right trailing by one mile. They were not. But it was still early.

"There's his beacon," Harry said over the interphone.

"I'll cut him off in the turn," Major Stevens said.

Ernest began hearing at least one of the gunners' radars. His receivers were tuned to listen to radar signals as if they were radio broadcasts. The gunner's radar sounded like a dog's growl. He spotted them and decided to wait. He didn't have an altimeter but could track their climb by time and the pacing of the copilot's checklists. After the copilot finished the "oxygen station check" with the crew, Ernest decided he could wait no longer.

"Pilot, EW," Ernest called on the interphone.

"Go."

"Are we in position?" It was a clear night, and Ernest suspected the pilots were flying visually off lead.

"Close enough for government work," Major Stevens said.

"Ah, not for much longer. We can't overpower a SAM's uplink solo. We need to maintain cell position."

"I know that. We still have time, don't worry about it."

Ernest sat in his seat, incensed that his life was in the hands of a pilot who knew less about how to fly the aircraft than he did. The ALT-22 was good, no doubt about it. It was far better than the ALT-6Bs on the newer B-52Gs and it was more reliable. But once they got within 10 miles of a Surface to Air Missile's fire-control radar, called "Fan Song," it wasn't good enough on its own. Three B-52s flying in a closer formation could defeat the Fan Song. But you had to be in position to do that.

"One hour to IP," Tommy announced. Ernest rechecked his clock. They had been airborne one hour and thirty-five minutes. The Initial Point on the bomb run was where all the fun really happened.

"Here comes the rest of our party," Harry said. "One rotating beacon after another." As the copilot, he would have the best view of the Andersen B-52s flying over South Vietnam to get into line behind the U-Tapao procession. "How long do you figure?" Harry asked over the interphone, to no one in particular.

"I've done that flight from Guam," Kevin said. "Twelve hours, plus or minus. It's a long way to fly to turn forests into toothpicks."

"And then they got to turn around and fly back," Tommy said. "Twenty-four hours of flying. That can't be easy."

"Can they do that without a tanker?" Harry asked.

"Sometimes," Kevin answered.

"How is that possible?"

"You are forgetting they are only carrying 27 bombs."

"How many we carrying?" Tommy asked.

"One hundred and eight," Kevin said.

"So why even bother sending a G?" Tommy asked.

"Good question that has absolutely nothing to do with what we are doing right now," Stevens said, ending all conversation.

Ernest heard and then felt a slight reduction to the thrust coming from half the engines. It was a Major Stevens technique. As they burned off fuel, he would pull back power on every other engine in increments of the width of the throttles' knobs. It was sloppy, but it got the job done. They were getting close.

"EW, Nav. One-twenty-five out," Tommy said.

"Roger," Ernest said. "Time to show the world where we are." He leaned his head into his radar receiver's scope hood and noted the other two aircraft were already testing each of their jammers. Ernest followed suit, that was what the rulebook said. He knew the North Vietnamese "Spoon Rest" radar operators would be able to locate the source of the jammers and would now have a pretty good idea where their targets were. But those were the rules.

While the test was in progress there wasn't much else to see, and the jamming signals had the effect of not only blinding the enemy but blinding the jammers too. Once the three aircraft were done, Ernest used his eyes and ears to see what was left.

"Ten minutes to IP," Kevin said.

"Get closer," Ernest said.

"What have you got?" Major Stevens asked.

"Nothing but search radars so far," Ernest said. "Get closer." A faint sound got his attention and he looked intently into his scope. The traces were slight. The Radar Warning Receiver showed no threats.

"Something," Ernest said. It was just like every training scenario back in school. Nothing begins overtly.

"Fan Song," they heard on the attack frequency. "Gold cell has Fan Song," the voice repeated. "Aborting."

"Amber Cell," the lead aircraft called, "Exit heading, ready, now." Ernest felt the aircraft bank to the right.

"Gold cell, secondary is buster."

"You heard the man, combat exit checklists," Harry announced as Major Stevens banked the aircraft to follow lead's retreat.

"I wonder what happened to the secondary," Major Stevens said. "Oh well, another day, another dollar."

"I just want to bomb something," Ernest said.

5. ROE

Omaha, Nebraska

"Explain the Rules of Engagement, Colonel," General Ham said. He was seated at the end of the conference table, his conference table on the top floor of the headquarters building. He was still new to his job as the Commander in Chief of the Strategic Air Command, but part of the art of being a four-star general was to play the role like you've always been there. His predecessor, a lifelong bomber pilot, was "asked" to retire by the President of the United States. Of course, when the President asks any question of the military, the answer is preordained. Now General Avery Ham had direct orders to fix what was broken.

He called a meeting of the CINC's Action Group – his action group – before addressing the headquarters staff. "Take all the time you need; you've got sixty seconds."

Colonel Snedeker was at the opposite end of the table and was flanked by his staff. Major Stephen Bellum, the junior man on that staff, sat at Snedeker's right. Snedeker cleared his throat. "General, every B-52 in theater has a primary commitment to the nuclear mission, it is why we as a command exist. We designed the ROE to prevent the loss of a single bomber, as they are no longer being made and are therefore irreplaceable."

"Okay," General Ham said. "I get that. But why do we have to abort so many missions?" The general was an imposing figure, even dressed only in his standard light blue shirt with dark slacks and four stars on each lapel. He chose not to wear the standard issue nametag. If anyone failed to recognize General Avery Ham, CINCSAC himself, on first sight, there would be bigger problems than a missing nametag.

"General, the primary threat is the SA-2."

"Which is something we learned to defeat years ago!"

"Yes, sir. Let me defer the answer to Major Bellum."

Stephen quietly cleared his throat. Even after meeting and speaking with ten general officers during his introductory week, speaking to the four star was an entirely different experience.

"General, the SA-2 has limited maneuverability during the first six seconds of flight," Stephen began. "After that, it can be steered by ground radar and an internal autopilot for 22 seconds. At that point it runs out of fuel and loses speed rapidly. A good fighter, provided it detects the missile first, can outmaneuver it to the point it has no fuel. The missile explodes at a preset altitude or time. Anything within 800 feet or so gets sprayed with shrapnel."

"Like I said," General Ham interrupted, "it is easily defeated."

"We just can't maneuver like a fighter. Our only real defense is to detect it and jam it."

"You ever been shot at, major?" the general asked without waiting for an answer. "Well, I have. It is a part of wearing the uniform."

"Yes, sir."

"Yes, sir it is a part of wearing the uniform or you've been shot at?"

"Both, general." The general slumped back in his chair, waiting for more. Stephen thought he needed to measure his words carefully. The President was clearly sending a message to SAC when he appointed a World War II era fighter ace to lead the nation's bomber command. The general could make or break any of his officer's careers with no recourse available.

"I've flown eighty-seven missions in theater, general," Stephen said. "It was mostly triple-A at first but this year we noticed more and more SA-2 activity. When we started moving the bomb line north, the mobile SA-2 launchers were deployed south to get us."

"I guess they damaged a few of our Buffs," General Ham said. "But no kills."

"Yes, general. We got lit up a few times, and in April we got fired on. But the ECM works."

"Exactly! Which is exactly why we need to change the ROE!" He didn't wait for a retort. "We would never put up with this weak ass ROE in the fighter community. I got a staff of bomber puke general officers here without the guts to fight a war. We're going to change the ROE." Turning his gaze to Colonel Snedeker he said, "Make it happen."

"Yes, general," Snedeker said. Satisfied, Ham got up and left without another word.

"That's going to bite us," Stephen said.

"Try to remember our job here, Stephen. Two jobs, actually. First, we represent the staff to the CINC. What they are thinking is what we have to convey to him. Second, we represent him to the rest of the staff. We give him the best advice we can, based on what the staff tells us. But once he's made a decision, it becomes the only thing that matters. Before the decision, we wholeheartedly agreed the ROE was sound and preventing the loss of even one B-52 was paramount. But once the CINC decides against that, that becomes our new position."

"It is the nature of being in the military."

"Exactly," Snedeker said.

"But what if the decision is so wrong, it becomes illegal, immoral, or unethical?"

"Let's hope we never get to that point."

6. Absolutely Nothing!

Berkeley, California

"What is war good for?" the professor said from the front of the class.

"Absolutely nothing!" most of the class yelled back. Martin thought immediately of Edwin Starr's song that became the anthem of many Vietnam War protests. In a way it was funny but hearing it in class from a college professor seemed odd. It was a large lecture hall, perhaps two hundred seats, mostly filled. Martin sat in the back row with Sherrie. He readily agreed to attend, thinking it would be their first date of sorts.

"Isn't he great?" Sherrie asked at one point during the lecture.

"He is," Martin said. But he didn't think that at all. The professor was ranting about the latest news, not really teaching any philosophy at all. Looking around the lecture hall, Martin noticed that half the students were paying rapt attention, half of the remainder were barely awake, and a few looked openly agitated. Finally, as the hour came to an end, the professor seemed to summarize his main points and ended how he started.

"What is war good for?" the professor yelled once more.

"Ending slavery," someone yelled from one side of the room. The professor was speechless.

"Ending fascism," another voice yelled.

"And imperialism," from a third voice. That was the straw that broke the camel's back, and half the class was on its feet as the other half sat in stunned silence. Martin was in the latter category until the first fist was thrown, at which point he grabbed Sherrie's hand and headed for the nearest exit.

Once they made it outdoors, they both started running instinctively, but slowed to a walk once the noise behind them had faded. "That was scary," she said at last.

"That was nuts," he agreed. "Grab a coffee?"

"Sure. Maybe you can help explain what that was all about." Their favorite coffee shop was just off campus but along the way to the dorms. It was closer to dinner time than coffee time and they both opted for a burger, fries, and a Coca Cola.

The front counter of the coffee shop was often packed, diners wanting to catch a view of the lone television hanging on the wall or knowing that the most comfortable seats were the swiveling stools bolted to the floor and neatly spaced along the counter. Martin preferred the booths, even though the seats were worn, and the tables etched with graffiti. He usually picked an empty booth and took his time reading whatever text he had. Having a guest with him was a rare event.

"Did war really end slavery, fascism, and those other -isms?" Sherrie asked.

"You could say that. But that would be a pretty myopic way of thinking about it."

"Myopic?"

"Short-sighted. History is filled with examples of a bad thing being ended by war. So, you could argue that it was World War II that ended the fascism of Hitler, Mussolini, and Tojo." He emphasized the word "could."

"So those guys had a point, war is good for ending those -isms."

"What morons like that don't get is that war is an awfully expensive and immoral means to getting to that end. Hitler and Mussolini were legally elected. Tojo was legally appointed. They were political creatures. Everyone thinks that Carl Von Clausewitz said 'war is politics by other means.' That's not what he said."

"Who is . . ."

"General Carl von Clausewitz. He was a Prussian general from the 19th century and his theories have shaped everything that followed. What he really said is 'war is the continuation of politics with the addition of another means.' And that makes all the difference."

Sherrie sat and looked at her half-eaten burger. She was a pretty girl, Martin thought, but a little dim. He had spent so much time keeping to himself that he hadn't dated since high school. He thought that perhaps her company was the answer to his loneliness but maybe not.

"Did war end slavery?" He asked, trying to explain it once more. "Yes, you can say that following the U.S. Civil War, slavery as an institution ended."

"So, it is a good thing then."

"No. That war bankrupted the country and killed nearly three percent of the population. War resulted after the politics gave up on fixing things without war."

"I see," she said. Martin realized that she did not. "Let's say the north just said, okay, south, go off on your own way. You are a separate country."

"What about the slaves?"

"A ha!" Martin said. "What about them? You cannot have a country survive on the cotton trade alone. Left on its own, the south was doomed to failure. We would have gotten the same result with less killing."

"Why hasn't anyone else thought of this?"

"They have, but the guys with the answers are never the guys controlling the levers of power. So that's why we have war."

"I worry about you going off to war," she said, finally. "I can't bear the thought."

"I do think about it a lot," Martin said. "When my draft number came up in 1970, I thought I might volunteer for the Air Force instead. My brothers are officers and I thought there was a lower chance of getting killed that way. Turns out Air Force officers are more likely to be flyers and flyers are more likely to be shot down and killed or captured. So, I ruled that out."

"Your brothers weren't worried about being shot down?"

"It was different back then. There was no draft, and hardly anyone even knew Vietnam existed. I thought I had no choice. But my dad came up with the idea for a college deferment."

"Your dad? I thought he was in the Air Force too."

"He was. He retired about that time. But I think he was worried about the war too. He helped me get the deferment."

"But deferrals come due."

"They do. June seems such a long time from now."

Martin walked Sherrie to her dormitory and said good night. He wrestled with the idea of leaving her with a kiss on a cheek, but did people do that anymore? As he crossed the quadrangle to his dorm, his mind raced with doubt. He had assumed the war would be over by the time he graduated, and that he could move on with his life. What kind of stigma would be put on him as a conscientious objector? Could he find a good job as a CO? His family would never forgive him. He checked his watch. His father would be at work; it was the perfect time to call.

"The Bellum residence, Mary speaking," she said. That was the proper way to answer the phone in the Bellum house.

"Mom."

"Martin? A middle of the week call? How nice. How is my future history professor doing?"

"The new draft numbers are out, and I was thinking about the war."

"I try not to do that. I am worried sick about Ernest."

"Ah, me too. How is he?"

"Well, you know Ernest. He doesn't like to talk on the phone, and he never talks about his feelings."

"Yeah, that's Ernest all right. Mom, I was wondering. How is it that he and Stephen didn't have any problems with joining the Air Force with a war going on?"

"I don't know. I suppose it is just what boys did back then. Before the draft, it didn't seem like a real war. Does that make sense?"

"I guess so. I feel sometimes like I came from a different family. I am so different from Stephen and Ernest."

"The three of you are like you each came from different molds. Stephen was a typical boy, playing cops and robbers, cowboys and Indians. When Jonathan came back from Europe, Stephen was all about airplanes. He had model airplanes . . ."

"Mom . . ."

". . . all over his room. We had no doubt that he would end up a pilot, just like Jonathan."

"Mom . . ."

"Well with Ernest, well that was different. He didn't care about any of that stuff. He was all about building things. I will never forget that when he was only two, he would take apart and put back together all the pens in the house. Can you believe that?"

"Mom, I've heard this before. Next you are going to tell me he built his own A.M. radio when he was six."

"He did! And then . . ."

"Mom, can we stick to the subject?"

"Oh, Martin. I'm sorry. What were we talking about?"

"Me, I think."

"You, of course. You were not like Stephen or Ernest. Stephen was 'Mister Social' and Ernest always kept to himself. You had your friends but mostly you liked to read about history. We knew you would major in it."

"I think I'm going to be nominated to stay on for a master's degree here. That is if the army doesn't get me first."

"The army? Jonathan thinks the draft will be over soon. The army is scary."

"War is scary."

"Flying airplanes, well it's different."

"Were you scared for dad, when he went off to war?"

"Of course. We had only been married a few years when he went off to training. It seemed those airplanes over Germany were dropping like flies. But by the time Jonathan got to the war, I think it had all changed."

"I know, I've taken a lot of courses on World War II. Dad couldn't have planned it better."

"I like to think we Bellums are lucky," she said. "Maybe Senator McGovern will win, and you won't have to worry anymore."

7. Brother's Keeper

U-Tapao, Thailand

Ernest showed up at the officer's club at precisely the same moment they had met that first night, but there was no sign of her. It was a stroke of luck that the table – their table – was free. He took his seat and looked around for a waitress, but there were none. So, he sat in his chair, waiting for *her*. He wondered if being in place was a sign of desperation and making her wait for him might be a better strategy. No, he ruled that out instantly. She was too ripe for the taking by any pilot on the prowl.

He heard the footfalls of a man, probably around one hundred and eighty pounds by the weight of each step and at least six feet tall, judging by the length of the stride. As the steps approached from behind and the person came into view, he looked up to see a major holding two Singhas.

"Ernest Bellum?" the major asked.

"Yes, sir," Ernest said, starting to rise until the major motioned him to remain seated.

"I'm Bob Colby," he said. "I come bearing a gift of Singha Beer and news from your brother." He was what Ernest considered a generic pilot. He was trim, didn't wear glasses, and spoke with an assumed authority that grated on all others, except other pilots. Like his brother, this major appeared young for his rank.

"I'll take the former," Ernest said, reaching for the beer. "The latter depends on the news."

"Only good news," Colby said, sitting in the opposite seat. *Her seat.* "He's doing well and says if you need anything, just let me know and I'll relay it. Stephen and I were classmates at the

Academy and squadron mates here at U-T. When he went off to work for General Ham, I went off to work for General Silvers."

"Who is?" Stephen asked.

"He's the 17th Air Division commander here at U-Tapao," Colby said. "Your squadron is under your wing which is under the air division. Then comes the 8th Air Force and then comes SAC." Colby tilted his Singha bottle and said, "Cheers." Ernest complied with a clink of the bottles and said, "Cheers."

The fact there were two levels of command between the wing and Headquarters SAC only confirmed to Ernest that the war was being run by a bureaucracy. He failed to see any good news in that.

"So, are you settled in?" Colby asked. "How is life as a combat EW?"

"What is the purpose of the 17th Air Division?" Ernest asked.

"Purpose?" Colby's face betrayed his surprise. "Purpose," he stated again. "Well, I suppose we are the conduit of command between the wings here and the 8th Air Force in Guam. We facilitate communications."

"The locals say Beer Singha. Not Singha Beer."

"What? Oh." After a few seconds he recovered his chain of thought. "Stephen speaks highly of you. He says they rewrote the Electronic Warfare book after you graduated from EW school."

"Why are we prohibited from using our chaff dispensers?"

"What?"

"You said you facilitate communications. I am communicating a problem with our ROE that can get someone killed. The ROE says we cannot under any circumstances dispense chaff from the B-52 when targeted by a SAM. Yet the fighters are not only allowed to do that, some have that as their primary purpose. Nobody at our level can explain why. We are left to think it's because our maintainers are too lazy to refill the dispensers. That's an implausible answer, but it's the answer we are left with. Maybe one of the many levels above can explain it to us in the trenches."

"Oh," Colby said. He appeared to be at a loss for words. "Let me see what I can find out."

Ernest turned his head to the entranceway and made eye contact. *With her.* She smiled and headed towards the table. *Their table.* Ernest looked to the major. "Anything else, sir?" The major appeared confused. Ernest got up, "I have a date."

"Oh," Colby said as she approached. "I'll get back to you with your question." He turned, made eye contact with her and simply said, "Lieutenant."

"Major," she said.

Ernest held the newly vacated chair for her. "I guess he was a friend of my brother. A spy in our midst. The least he could have done is brought another beer," Ernest said, spotting a waiter. "We'll fix that right now."

"Thank you," she said. "I'm rather new to this military officer thing. Do I call you captain, or can I call you Ernest?"

"Ernest, of course," he said. He paused. "I'm kind of embarrassed to say this," he said. "But I am not very good at this. I don't even know your name. Other than the name on your uniform. I don't want to call you Lieutenant Ralston."

She smiled. "It's Marjorie. Everyone at the weather shop calls me lieutenant. It will be nice to hear my name again."

"Marjorie it is."

"Pleased to meet you again, Ernest. Now that I've been here a week, I kind of like it. Other than the fact we are in the middle of a war, it almost seems like a good place for a vacation resort."

"I do believe there are vacation villas in the area. I think the Germans particularly like Pattaya Beach. *Das Leben ist schön!"*

"Es ist so," she replied.

"You are full of surprises, Marjorie."

"It now makes sense. You have a very good ear for sounds, so it makes sense that you are multi-lingual."

"Not really. Just English and German. And I guess EW. Okay, I am trilingual. How many languages do you speak?"

"One less than you. And I suspect your German is probably better than mine."

"We lived in Germany for a few years. Languages are easier when you are a kid."

"Did your father fly in Germany after World War II?"

"No, he was a staffer. I don't think he ever flew after the war. After his war. He spent the rest of his career on the staff. Kind of like that major who is spying on me. My dad was a facilitator of communications. Just what this war needs, more people talking and less people doing."

"I know what you mean. Our top meteorologist here is a captain with less than six years on active duty. There are twice as many forecasters at the headquarters than here in the field."

Ernest nodded his head as she spoke; it was a common refrain from just about every corner at U-Tapao. "Rank has its privileges, I guess. My dad was a career man. Stephen too."

"What about you?" she asked.

"I doubt it. I still have connections at MIT. I owe Uncle SAM four years for the training. After that? I will probably punch out."

A cool breeze wafted into the covered patio, easing the oppressive heat, and allowing everyone inside to collectively inhale a change of temperature and perhaps mood. Ernest noticed for the first time that Marjorie's red lipstick complimented her brown eyes and her auburn hair. The thought made him smile. "How about you? What's next?"

"I come from a family of musicians but have always preferred math to violins. My dad is the Director of Music at the University of Oklahoma, my mother taught there, and my older sister is on the New York Philharmonic."

"The big leagues."

"It is. I started as a music major but was unhappy with all the practice. I wanted to change to math, but my dad said I would never make a living doing that. One thing led to another, and meteorology seemed like the best compromise. Not very interesting, I'm afraid."

"Do you still play?"

"Yes. It isn't that I don't like playing; I love playing. But as a hobby, not having to play the same thing over and over again, trying to compete with others leading up to 'first chair' or maybe someday becoming a conductor. I gave lessons for a while but that became tedious too."

"Do you still have your violin?"

"Yes, but I didn't bring it with me."

"What kind is it?"

"Oh, you will have never heard of it, it is a Benning."

"Nice. Did you get that stateside or in Germany?"

"In Germany, it was a family trip. You know violins?"

"I spent some time in Mittenwald, Germany right after EW school. I think the Bennings learned their craft there. I was impressed with the craftsmanship and I do enjoy violin music. So, I bought one."

"You bought a Benning in Mittenwald?"

"No, I bought a Mittenwald in Mittenwald."

"Wasn't that expensive?"

"Very. Five grand, I think it was."

"I would love to hear you play."

"I'm not very good."

"Oh, you are just saying that."

"No, I'm not. I've never had a lesson and am trying to teach myself. Maybe I can hire you as my instructor."

"You will be my first student since college."

"Perfect."

8. Press On!

Omaha, Nebraska

General Ham unveiled his new orders at his weekly staff meeting, attended by every division director on the headquarters staff. Everyone at the table was a brigadier general or higher. The seats orbiting the table were reserved for supporting staff, including two from CINCSAC's action group: Colonel Snedeker and Major Bellum. It was as regal a setting as Stephen had ever seen in the military. The large conference table was surrounded by padded, high-back chairs. There were paintings of airplanes and ballistic missiles on the curtained walls, but no other furniture. All officers were in their Class A uniforms, a light blue, long sleeved shirt with tie, topped with a dark blue coat with rank on the epaulets and ribbons over the left pocket. Most of the generals also wore aviator wings or a badge displaying a missile.

"I got this directly from the President," General Ham said, "so I don't want to hear any debate. This is the way it is going to be. If our crews are faced with a surface-to-air or even an air-to-air threat, they are to press on to their targets. There will be no more turning tail at the first sign of trouble. I can sum this up in two words: press on!"

Stephen had the advantage of hearing the new ROE beforehand, but hearing it articulated so clearly was chilling. By the looks on all but one of the subordinate generals, it appeared they had not anticipated the change. The lone exception was Brigadier General Mayfield, the head of the Electronic Warfare Division. He appeared almost giddy with excitement. General Ham finished his prepared remarks and looked up from his notes.

"How do you plan on implementing the new ROE?" He turned to his right. "Alan?"

Lieutenant General Alan Stringfellow was the Director of Operations, ostensibly the number two man in the command and the leader of the SAC Operations Staff. He managed a smile and sat up in his seat, to CINCSAC's right. "We will have the entire operations staff go over each of our existing battle plans and, line-by-line, ensure the new ROE isn't conflicted by other procedures. I'll personally give 8th Air Force a call to make sure they are on board."

General Ham next turned to the Director of Intelligence, Brigadier General Michael Ferrell. "Mike, how's all this going to playout for Linebacker?"

"Unclear, sir," Ferrell said. "The North Vietnamese have been busy positioning mobile SA-2s throughout southern North Vietnam all the way up to Hanoi and beyond. Their SA-2 regiments have been somewhat, shall we say, ineffective?"

"That's an understatement!" Stringfellow said, chuckling. Laughter rippled around the room.

"But that won't last for long," Ferrell continued. "We believe the Soviets sent a hundred SA-2 experts to North Vietnam in the last few months. We have increased SIGINT to suggest they are practicing for the next time we attack Hanoi."

He gestured to the orderlies at the overhead projector. "Slide seven from my stack," he said. The screen filled with a summary of Signals Intelligence intercepted by RC-135 spy planes orbiting east of Hanoi over the Gulf of Tonkin. "Where you see 'S-75' think of the SA-2. S-75 is what they call it. As you can see, mentions of the S-75 Dvinia have gone up several hundred percent in the last six months. But more troubling is we believe about a hundred Soviet S-75 advisors have been in country."

"Teaching them how to operate the thing?"

"Yes, general. But more than that. Perhaps helping them to adapt the missile launch systems to our jamming. The S-75 has done well against our fighters."

"Can the ECM on our B-52s handle this?" General Ham asked, looking at Brigadier General Mayfield.

"Yes, general they can," Mayfield said. "As we found out in April, the North Vietnamese have a lot of Soviet hardware that SAC has been studying for years. We know how to defeat the SA-2."

"Excellent!" General Ham said. "Let's get to it." With that, he rose, and the staff stood at attention. While standing, Stephen watched Brigadier General Ferrell, who appeared to be on the verge of speaking just prior to General Ham's "Excellent!"

"Did you catch that?" Snedeker asked Stephen.

"Yes sir, he was about to add something."

"Find out what he was worried about but unwilling to say."

Stephen took off, hoping to catch up with Ferrell but he was too late. The door at the Intel Division's office closed shut just as he arrived. He picked up the phone at the door, was met by a sergeant, and was then granted access to the vault within.

It was a typical intel vault, much like at any SAC base. There were no windows, the walls were covered with maps of various parts of the world, mostly inside the Soviet Union. But unlike most of the stateside intel vaults, this one had a wall that was dedicated to North and South Vietnam and included a broader map that included the B-52 bases in Guam and Thailand.

Brigadier General Ferrell was seated at one of the desks, staring at a large map of Hanoi. The orderly got his attention and Ferrell looked up. "Major Bellum. I gather you have either CINCSAC questions or CINCSAC answers."

"Neither, general," Stephen said. "I noticed you were about to say something when General Ham dismissed us. I just want to make sure any concerns or issues that you have are addressed."

"Oh. I guess that means we aren't in any trouble, I suppose." He looked visibly relieved and offered a weak smile. "Come into my office and we can talk."

"Concerns or issues," Ferrell said, as he sat at his desk, looking above Stephen's head, placing his focus at infinity. Stephen noticed a photo on the wall of President Nixon pinning a medal on

Ferrell's chest. Then another with President Johnson. "I wouldn't say I have any issues," Ferrell finally continued. "The ROE says we attack and that is why we are there. Concerns? I guess I have a few."

"I suppose that is part of the job description, general."

"It is indeed. I wonder if we are underestimating our enemy's capabilities and, at the same time, overestimating ours. Since Operation Linebacker began in May, we've had one airplane damaged by a SAM and that's it."

"Pretty good report card."

"Is it?" Ferrell asked. "We, and by that, I mean the United States, have lost over a hundred aircraft. All of those were fighters, about fifty-fifty between us and the Navy. That's not so good. The North Vietnamese claim thirty of those fell to the SA-2."

"But not the B-52."

"No, not the B-52. But ever since we got dusted up in April, we haven't been sticking around for the fight. And in that time, the bad guy has been studying us, and adapting. We haven't been doing the same."

"And now we are back in the fight."

"Exactly."

9. The Question

Hanoi, North Vietnam

"Engineer Major Yakolev," he heard from the door to his makeshift office, "we have the radar photos you asked for."

"At last," he said, gesturing the North Vietnamese Army major in. The NVA officer was dressed in short sleeves and pants and appeared to be comfortable with the heat and humidity.

"The Soviet Army no longer uses functional titles. You can call me Major Yakolev if you like. But as we are both majors and allies, I would prefer you call me Nikolai."

"Of course, Nikolai. The photos do not have identification marks, I shall place them in the order they were taken."

"This was from the attacks in April?"

"Yes, Nikolai. The center of each photo is from the radar site you visited yesterday. Allow me to label them for you."

Nikolai stepped back from the desk so the missile battery commander could write on the margin of each photo. The three photos were remarkably different. "Do you have a time reference?" he asked.

"It was chaotic, as you can imagine. But we think the first photo is from when the acquisition radar had activated, one minute prior to missile launch. The second was six seconds after launch, after we successfully uplinked to the missile. The last, is when the missile detonated."

"A miss, of course."

"Of course. The missile reached the target altitude, but it looks like the bombers turned to the right."

"To the right?"

"To the aircraft's right, Nikolai. The scope is oriented north-up, the aircraft are approaching from the north."

"Ah yes, I had forgotten that bit of American madness. They approach from the north and turn to the right."

"That makes it easier for us," the North Vietnamese major said, smiling. "I must be off, but do not hesitate to call me back."

"Yes, of course," Nikolai said. He studied the photos, starting with the first. The time difference was no more than two minutes from first to last, but they were completely different. Nikolai caught a bead of sweat from his brow hitting the table, just missing the photo. He pulled back and wiped his forehead. Even in October, the heat was oppressive. The photos were invaluable. There was another knock at the door.

"I have those towels you asked for, major," the Soviet captain said. His uniform was more sweat stained than Nikolai's, but he had been outdoors all morning.

"Thank you, Sergei. How has your inspection gone this morning?"

"I must admit I was skeptical, but the heat and humidity does explain some of the difficulties. I have impressed upon our allies the need to keep everything clean and to refrain from using the electronics without the cooling systems operating."

"Very good. What of the mathematics we spoke of earlier?"

"It is a simple matter, major. But there are many variables. We need to measure the distances and bearings between each search radar, triangulate the targets, then triangulate the radars themselves."

"If it is simple, what is the problem?"

"Many of the components are moved frequently, to prevent the Americans from learning their location. But even with the stationary stations, the measurements take time."

"How much time?"

"We shall have the measurements in a week. But even once that is done, it will take time. The remote sites must spot the bombers, do the calculations, send the numbers to the central control agency. Then the missile battalion commanders must do their own calculations. The errors compound and if they are not precise enough, the missiles do not have a chance of acquiring their targets. We will conduct drills. I feel confidence, they shall learn the process."

"And we too shall learn. We can bring these methods home and apply them for ourselves," Nikolai said with the first bit of optimism he had felt since their arrival. "What do you make of these radar photos? The bombers are approaching from the top. This is before the missile is launched, this is at the moment of uplink, and this is the self-detonation."

The captain studied each photo. Major Nikolai Yakolev was their team's leader. But the captain was a trained SA-75 commander.

"The thin spikes radiating from the center to the outside of the first two photographs are the acquisition radar signals," the captain said. "They have acquired the target on the first photo and also the SA-75 on the second."

"And the wider spikes?"

"The wider spikes on the second two photographs are the uplink signal to the missile."

"They had a good uplink? Why, then, did the missile miss?"

"Difficult to say from these photos, but it appears the initial direction was not as accurate as it should have been. The missile may have simply run out of fuel by the time it got close enough. The light areas that are especially visible on the first two photographs are the enemy noise, the jamming of our radar and the uplink. It could be the jamming of the acquisition radar was too powerful."

"Where are the bombers?"

"You can see at least two of them on the last photo. They are the distinct dots. It appears they are turning."

"But where are they in the first two photos?"

"Inside those clouds of noise will be the bombers themselves. The jamming is quite severe, but it is very much reduced in the third photo. Looking at the third photo, behind the two dots, we have a smaller cloud of noise. There is a third bomber we do not see. So, there is a question, then."

"Yes, there is a question, indeed." The captain left with more towels to deliver, leaving Nikolai with his question. Why did the Americans decrease the jamming after the missiles were launched, just when they needed the jamming the most?

10. Nagged

Berkeley, California

"This is very good," Sherrie said. "It not only gets into the details of that war, but it reveals the lie of it."

"The lie of daylight precision bombing, yes," Martin said. "But not of the war itself."

"They are the same thing."

"That is a false equivalence. Daylight precision bombing was a tool used to win World War II. Daylight precision bombing was a mistake. But that doesn't mean the war was a mistake."

"I thought you said war happens when politics fails."

"Yes."

"So, we should not have gotten into World War Two then."

"I'm not saying that World War Two was wrong. I am saying that prosecuting the war using bombers of that era to go after the targets they had in broad daylight was a mistake. You cannot send hundreds of bombers with thousands of crewmembers into a nest of fighters and flak with so little probability of knocking out their targets. The strategy was wrong."

"All wars are wrong," she said, crossing her arms to emphasize the point.

"It's complicated," he said. It was an issue he thought he had settled in his mind. His term paper was focused on the war in Europe. He had started with addressing the American response to the Japanese bombing of Pearl Harbor but his conclusions didn't add up. It was complicated.

"This paper took me all semester to write. I wanted to show that sending our bombers over Germany in broad daylight, with the technology they had, was suicidal. That was my main point."

"But you wrote that right after the war, historians said daylight precision bombing was crucial to winning it," she said.

"But they got it wrong, which is really my entire point. They say history is written by the victors and the victors can make themselves the heroes and everyone else, well, not the heroes."

"So, what changed?"

"We got bomb damage assessments that showed this factory destroyed or that Army division wiped out. It took having a look at the enemy's records to learn the truth. A photo from 20,000 feet shows a ball bearing plant leveled, so we believe we succeeded. But looking at German photos we see the heavy machinery survived. All they had to do was relocate it and production continued. Time after time our first assessments were wrong. We know that now, with the benefit of time."

"There are lessons for Vietnam then," she said. "Just because our government says we are winning and 'peace is at hand,' doesn't make it so."

"True. But the North Vietnamese just agreed to return to the peace talks. Maybe the bombing has had some effect and this war is different."

"War is war," she said. "Why can't we go back to the way it was before? I wish I was Patty Duke."

"Who?"

"Martin, you don't know who Patty Duke is? She played identical cousins on TV not too long ago. She was cute, smart, and her biggest worry was what to wear and who to date. Life was simpler back then. If you would just submit your conscientious objector letter we could get back to normal. I am tired of worrying about the war."

"I know, me too. But . . ."

"But you don't want to upset the family," she interrupted. "I'm tired of hearing that. You are going to have to do this, and soon. You know that!"

"I know that," he admitted. "I'll submit it as soon as the check clears on my next tuition payment."

"Who is paying your tuition?"

"My dad. He saved up as best he could for all three of us. It wasn't going to be enough, but Stephen went to the Air Force Academy which cost the family nothing. Ernest was on a full ride scholarship at MIT, so I think my dad only had to pay for some of his living expenses. So, I got lucky. Dad said he would pay for all four years and after that, I'm on my own."

"So, one more tuition check and you are done with your dad?"

"Yes, financially speaking anyway. The day that happens, I'll submit my CO letter. I promise."

"I hope it is soon enough, Martin."

He realized, for the first time, that he was being nagged. She was a pretty girl with a simplistic view of the world. Maybe he was letting the pretty part make up for the simplistic part. But they were now a couple, he decided. He wasn't sure what that meant in the 1970s, but it meant something.

11. Olive 02

U-Tapao, Thailand

"Just a reminder, boys and girls," Major Stevens announced as he turned the bomber east, "we got a few changes to our ROE. First, there is no turning back if we encounter fighters or missiles. Second, we will be allowed some maneuvering prior to the target, but we have to be on course and level when we get to the target. Any questions?"

"If I get me a fighter today," Polanshek asked, "can we paint its silhouette on the fuselage?"

"If you get you a fighter today," Stevens answered, "I'll personally paint it."

It was Ernest's first mission north of the DMZ and though he was a little nervous, he felt confident.

"Ten minutes to IP," Kevin announced. Crossing Laos would only take a portion of those ten minutes and then their target, railyards outside Vinh, North Vietnam, lay in wait. His 125-mile jammer check was done, he had them all silent now and closed his eyes as he listened to his Radar Warning Receiver. The RWR scope would show strobes and its warning lights, such as the one marked "Launch," would illuminate when the time came. But it would also turn those signals into audio and his ears would know before the scopes or the lights.

"E-dub, pilot," Major Stevens called.

"EW," Ernest replied.

"No sign of the weasel. They were supposed to be here to take out the SAMs. What can we expect?"

"Intel says they have 200 SA-2 launchers and ten times as many missiles," Ernest said.

"Where?"

"Don't know. They are mobile launchers. It takes about a day to set them up, but they could be anywhere."

"Can you detect them?"

"We're late. They're probably gone by now."

"What?"

"The weasels," Ernest said. "We took off twenty minutes late, the weasels probably didn't have the fuel to stick around."

"Terrific," Stevens said.

"Two minutes to IP," Kevin said. "First aircraft should be over target now."

"I see 'em," Harry said. "Quite the show. Even at 36,000 feet, the bombs on the surface are lighting up the night."

Ernest rechecked his radar receivers. Nothing but search radars and the signals from the cells in front. The SA-2 was a dumb missile on initial launch until the fire-control radar locked on to its beacon and the radar signature of its target. The "Fan Song" fire-control radar had a narrow beam width, only 7-1/2 by 2-1/2 degrees. But Ernest's gear should be able to pick it up. But he didn't see any Fan Song at all.

"IP," Kevin said. "PDI, please."

"Got it, turning two degrees left," Stevens said. From the Initial Point on they no longer navigated as a cell. Instead of maintaining relative position to lead using his radar, Stevens would turn to center his Pilot Director Indicator, an unmarked compass card controlled by the radar navigator. If he wanted the pilot to turn, he simply rotated the card from his panel downstairs. The pilot, from upstairs, would turn the aircraft in response.

Since his RWR was silent and he had no other threats on his receivers, Ernest sat silently while keeping his head buried in the scope.

"Sixty seconds," Kevin said. "Doors."

Hurry up, hurry up, Ernest said to himself, imagining the enemy's radar view of them growing as the bomb bay doors hanging out in the wind announced to the bad guys: look at me!

"Thirty seconds," Kevin said.

Hurry up! Ernest said to himself again.

"Five seconds," Kevin said. "Bombs away." He counted down the bombs from 100 by tens as the aircraft leapt upward a hundred feet as 108 750-lb. bombs left the grips of their racks and the aircraft lost 40 tons of weight in just 15 seconds. "Release complete, standby."

Hurry up! Ernest said to himself for what he knew would be a very long 30 seconds as a sitting duck with those bomb bay doors open.

"Doors closed, clear to maneuver," Kevin said. Stevens turned sharply to the right in a 60-degree bank, inflicted forces twice that of normal gravity on the aircraft and everyone in it.

"Strobes, it's a SAM," Ernest yelled, even as he struggled with the weight of his oxygen mask contorting his face and his helmet trying to slam his head against his instrument panel. Unlike every simulation in training, the rattling in his headset became distracting. The "Launch" capsule on his RWR lit up. He lifted a heavy arm to one of his AN/ALT-22 noise jammers. "Launch! Jamming them."

"Tally ho, one o'clock," Harry said. "Headed right at us."

"Reverse your turn!" Ernest yelled.

"No," Stevens said. "We turn right."

"That ain't gonna do it," Harry said.

"Well, I'll fix this," Stevens said, increasing his bank and pulling back on his yoke harder. The G-forces doubled. Ernest's left arm, adjusting a jammer, was slammed to the table. The entire airplane shuddered in complaint.

"Stall!" Harry said. Stevens instinctively leveled the aircraft's wings and pushed the nose down. Ernest was pushed up into his straps, becoming weightless for a second and managing to adjust the last jammer on the latest frequency.

"Missile's gone ballistic," Ernest said, his voice lowering several octaves.

"I got it," the gunner said from the tail of the bomber. "Heading where we ain't." A few seconds later, "Detonated. Good job, Captain Bellum."

"Combat exit checklist," Broadmore said. After each crewmember called back to verify their checklists were completed, the crew fell silent until over Thailand again.

"E-dub, pilot," Stevens said.

"Pilot, EW," Ernest said.

"What did you mean the missile went ballistic?"

"The SA-2 is unguided until the ground Fan Song radar gets its beacon and then steers it," Ernest said. "I saw the uplink on my scope. If we jam it long enough, the missile will lose the Fan Song radar and it won't know where to go. They keep sending it instructions on different frequencies and we keep jamming the new frequencies. If we succeed, the missile becomes a ballistic projectile. Nothing more than a very big bullet. They can make it self-detonate after a time interval. At that point, it is just like flak."

"You did that?"

"I think so. I put every jammer I got on the uplink frequency. They changed five times in just those few seconds. I think I got them all. If that worked, it could have caused the missile to break lock and go ballistic."

"You don't sound too happy about it," Harry said. "You just saved our lives. If you can't pat yourself on the back, I'll unstrap and come back and do that for you."

"No, I get that," Ernest said. "But I don't understand why I couldn't detect the missile until after the uplink began. I should have known as soon as the Fan Song got us acquired. Something doesn't make sense."

The crew fell silent. Ernest sat and glared at his panel, accusing it with his eyes for letting him down. He never saw the Fan Song until it was too late. His noise jamming was a last minute, frantic, desperation move. The Old Crows described it as an "assholes and elbows" moment. It wasn't a moment he wanted to relive. The quiet lasted no more than a minute.

"Olive 2's been hit," they heard on frequency. "She's going down."

"Who was on Olive 2?" someone asked.

"That's Wheelock's crew," another answered.

"Oh."

"His gunner is that new kid, Harvey Something," Sergeant Polanshek said. "I think this was his first combat mission."

After their mission debrief it was well past midnight and Stevens decided their usual beer call could wait a day.

12. Pressed

Omaha, Nebraska

"What in the hell happened?" General Ham asked. "Take all the time you need; you've got thirty seconds."

Colonel Snedeker stood at attention in front of the general's desk. Stephen, standing to the Colonel's right, had grown used to the routine, though it still made him nervous. It was his first time in front of what the staff called "The Carrier Deck." The general's desk measured 120 inches wide by 60 inches deep, precisely double that of the standard issue variety.

"Olive 2 was the second airplane in the fifth cell attacking Vinh, North Vietnam, sir," he began. "She didn't see any radar activity prior to bomb release and was able to drop on target. They did detect the SAMs during their post target turn, but the missiles appeared to have lost lock at some point. But the missiles self-detonated pretty close by, and the shrapnel hit the aft fuselage and one of the wings, causing fuel leaks and fires. The pilot managed to keep her flying until she ran out of fuel, he then glided her across the Laotian border. Everyone bailed out around 15,000 feet. They were all picked up by a CH-53, no serious injuries."

"No!" General Ham said. "I know what happened. I want to know why it happened."

"Major Bellum has investigated that," Colonel Snedeker said, gesturing to his right.

"Of the eighteen airplanes," Stephen began, "only three detected any SAM activity at all. It appears the enemy has modified their tactics to leave the Fan Song fire-control radar off until after the bombers are overhead their targets. We are looking into ways to overcome this. What we don't understand is why our noise jammers don't appear to be working with complete reliability."

"Or at all?" the general asked.

"They do seem to be ineffective at times," Stephen said. "But in two of the three cases, the EWs reported that the missiles were successfully jammed at the last moment. We think that also was the case with Olive 02. But their missile detonated close enough to cost us the airplane.

"The ECM gear is perfect," the General said, "if you believe the guys who tested it. I'm not sure I do."

"It seemed to be perfect until yesterday," Stephen said. He started to say but stopped short that Olive 02 was the first B-52 ever lost in combat. Telling a general officer what he already knew was not a good idea.

"The crows in this building don't have a clue what's going on in the field," the general said. "They tell me Olive Zero Two was a fluke. But you are telling me we could have lost three airplanes last night. Do I have that right?"

"Yes, general," Stephen said.

"I'm not sure we can jump to that . . ." Snedeker started to say.

"I didn't ask you," General Ham said.

"Do you have confidence in the ECM, major?" the general asked.

"It worked for me," Stephen said. "But the enemy seems to have learned a way around our tactics."

"Exactly," the general said. He sat back in his chair. "At ease, gentlemen." From his peripheral vision, Stephen noticed that Colonel Snedeker relaxed his spine, spread his legs, and placed both hands behind his back. Stephen did the same. It was just like being back at the Academy, he thought.

"Those old crows are a strange breed," the general said. "Even in fighters. Half the time I don't have a clue what they are talking about and sometimes I wonder if they have a clue either. You ever spend any time with one of those, what do you call them, you know, the guys with brains on another level? Really smart in one way but dumb in another?"

"Idiot savants," Snedeker offered.

"Exactly," the general said. "I get the feeling these old crows have computer brains but lack the basic ability to speak to those of us with normal brains." He sat back in his chair, staring out the window. Stephen shifted his legs slightly, willing the blood flow to keep him upright.

"Major," the general finally said. "Do some undercover work. Find an EW you trust back at U-Tapao or Andersen. Find out what he knows and what he thinks will fix this mess. We cannot lose any more airplanes."

"This is going to get worse before it gets better," the general continued. "We are heading downtown."

Downtown. That meant Hanoi. A chill ran down Stephen's spine. He had been downtown back in April.

13. Collateral Damage

Berkeley, California

"I heard it was a drunk driver," Martin said. The phone connection was a good one, the operator cheerfully waited as he fed his quarters into the payphone and the hospital front desk transferred his call directly to his father's room.

"A drunk driver without a license," Jonathan said. "Funny how all those years flying in combat and the Nazis couldn't lay a finger on me. And now some drunken S.O.B. did this."

"Aren't you mad about this?"

"Everything happens for a reason. Sometimes we don't see the reasons immediately, or maybe ever. But getting mad won't help."

Martin had heard the refrain from his father before, but never in the context of being brutally maimed. He shifted uneasily on his feet, standing next to the hallway payphone, aware of the silence. "Does it hurt?" he finally asked.

"Hurt like hell when it happened. Right now, it just itches."

"Itches?"

"Yeah, my foot. The foot I no longer have feels like it itches. Funny." Martin smiled at the stoic bravery. He started shifting his weight between his legs, keeping the phone cord taut as he bounced side-to-side. It was something his father could no longer do, he suddenly realized.

"How are you doing, son? I imagine you are unhappy about the election."

"I was hoping for McGovern. He said he was going to end the draft and I thought that would keep me out of the Army when I graduate."

"I knew him, McGovern. We went to primary flight school together, back in Muskogee. He went to B-24s and I went to B-17s and never saw each other again. Something screwed him up in the head, I think."

"And now we are stuck in Vietnam. And me in the Army."

Martin could hear his father's sigh over the phone. He knew the senior Bellum usually only spoke after composing his sentences. Measuring words. Always.

"The war is winding down. Nixon may have won the election, but he lost the support in Congress he needed to keep fighting. The American public doesn't support it and the military works for the American public. You might end up in the Army, but it won't be for long and I bet you never go overseas unless you want to."

"I don't want to go into the army at all."

"I know you don't, son."

Martin pulled the phone from his head, looking at it in disbelief.

"You do?"

"Of course, I do. Don't you remember it was me who got the ball rolling on your deferment."

Martin's face felt flush, realizing he was wrong.

"I remember. I just thought you would eventually push me into joining the Air Force."

"I always knew Stephen was headed that way. I was surprised by Ernest. But I guess saying Ernest surprised us at anything is no surprise at all. We encouraged him to apply for every college with courses that interested him and when the Massachusetts Institute of Technology accepted him, well we just couldn't afford it. So, he applied for and got an ROTC scholarship there." Jonathan laughed. "Do you know he outscored Stephen on his pilot test? Where in the hell did he get that brain of his? And then he ends up an Electronic Warfare officer. Go figure."

"Everything happens for a reason."

"It does indeed. I don't think you are ending up in the Army. I have a gut feeling about this. All my boys are going to be fine."

"I thought all your boys were going to be in the Air Force."

"No, I never thought that. I thought Ernest would end up in some think tank solving the unsolvable problems, curing cancer, that kind of thing. I always thought you would end up as a college professor. I still do."

14. Damage Assessment

U-Tapao, Thailand

The SAC official pre-mission meeting took place the day before the mission and the official post-mission debriefing happened immediately after the mission. Major Walter Stevens usually added to the schedule with "mandatory beer call" amongst the five officers of the crew and after their first mission north of the DMZ, he set a beer call for the next night at the Officer's Club. Only the gunner was spared and only because of his rank. Ernest usually timed his arrival to be as late as possible, finding the comradery between the two pilots and between the two navigators to be a bit tiresome. But for this beer call, he was anxious for company, even theirs. He showed up as the bar opened, sat at an empty table, and ordered a beer.

He had had a fitful night's sleep and a restless afternoon. *This wasn't supposed to be like this.* Sitting, watching the condensation collect around the beer bottle did little to settle his thoughts until he heard the footfalls: lighter than a male's, weight around 120, the stride indicated about five-foot-five. He looked up. *Her.*

"I am so happy to see you," she said. "I was afraid you were shot down last night. The base isn't releasing any names."

"The crew is okay. They punched out over Laos and got picked up. Nobody got seriously hurt. Except the airplane."

"That's a relief. You were flying in the same group, though."

"We were two cells ahead of them. We heard it on the radio but didn't see it."

"Was it bad?"

"Kind of, but I can't talk about it. I am happy to see you too."

"Really?"

"You are the closest thing to a friend I have around here," he said.

"Oh, I doubt that. But I am happy to be counted as your friend. It can get lonely at a new base in a small weather shop where everyone seems to have someplace else to be."

"Just give me a call if you want company," Ernest said.

"I will," she said, pausing to look behind him. "I don't want to ruin the mood, but I think your pilots just spotted you. They are walking this way."

Ernest looked up just as Major Stevens spoke. "Good evening, everyone. Don't let me interrupt anything, I just wanted to buy my favorite EW and his date a beer," he said, placing two cold bottles of Singha on the table. "Ernest, you carry on, I think we debriefed this one to death, don't you think?"

"Yes, sir," Ernest said. "Thanks."

Harry slapped Ernest on the back and looked at Marjorie. "You take care of our boy, here. He took care of us last night."

"Okay, now you have to tell me!" she said as soon as the two pilots were out of earshot. "What happened last night?"

Ernest quietly considered his beer and his next words. Much of what he knew suffered from two problems: it was top secret and it was speculation on his part. Either way, if she repeated something she wasn't supposed to know, they would both be in trouble.

"My job on the Buff is to keep the bad guys off of us so that we can drop bombs on them. And once we do that, my job is to get us safely home."

"The beeps and the squeaks?"

"Yes," Ernest said, delighted that she remembered. "It is a chess game where you can't see the other player's pieces. Our pieces are designed by people who only have knowledge of the game as it has been played in the past. But the game is constantly changing. Does that make sense?"

She nodded. "I read about electronic counter measures in the wing history they gave us during our indoctrination. I take it the big threat comes from SAMs and they use their electronics to aim their missiles, and you use your electronics to fool their missiles into going someplace else."

"Couldn't have said it better myself," Ernest said, raising a glass. "We should make you an honorary EW."

"No thanks. I get airsick just watching you guys takeoff. Besides, I think you guys are going to get even busier."

"What makes you say that?"

"The head of the weather service is going to beef up our shop here at U-T. They are asking for ten volunteers stateside to deploy here in the next two weeks. And if they don't get volunteers, they are going to get volunteered."

Ernest thought that perhaps he had just learned something he wasn't authorized to know, instead of the other way around. From the corner of his eye, he saw Major Stevens approaching rapidly. Perhaps being left alone was too much to ask.

"Ernest," he said with a nod to Marjorie, "I got a call from the command post that says you need to double time your way there. Believe it or not, you have a call from CINCSAC's office."

"I guess I'd better go then," Ernest said. He suspected it was his brother but left that out for dramatic effect. "Marjorie, I hope to see you tomorrow?"

"Of course," she said.

15. An Inside Source

Omaha, Nebraska

"Please give this to CINCSAC," Stephen said to the secretary. "He's expecting it. I'll be in my office." That chore done, Stephen strolled the 50-yard distance to his own office, opened the door and was flagged down by his own secretary just cradling the phone.

"Major Bellum, you saved me the trouble of writing you a note," she said. "That was CINCSAC's secretary. Her message from the general to you is, 'get in here now!' So why are you wasting your time talking to me?"

Stephen reversed course and this time CINCSAC's secretary said, "Walk right in, major."

"Start talking," General Ham growled.

"General, my brother is an EW at U-T," Stephen said, resisting the urge to loosen his tie. "He flew the same mission where we lost Olive Zero Two, in fact he was two cells ahead and listened to them as they got hit and struggled over the border."

"Did he see any SAMs?"

"He saw them on his radar receivers. His copilot and gunner spotted them visually."

"But they didn't get hit."

"No, sir," Stephen said. "But each airplane in their cell went through the same gyrations. It sounds like half the airplanes between his airplane and Olive Zero Two had similar troubles. He thinks they all got lucky, except Olive Zero Two."

"So, it is his considered opinion that the fate of our B-52s yesterday was determined by luck? The Electronic Warfare

Division here says everything worked perfectly. Who is right and who is wrong?"

"Maybe a little of both, sir. Ernest says the jammers worked perfectly, but he just barely had enough time to detect the missile and do his EW magic."

"So, he's the problem?"

"No, sir. He not only graduated top in his EW class, but he was the top graduate they ever had. Besides, the other EWs reported the same issue. And half of those EWs were veterans from the SAM launches back in April."

"Why, then, were they so rushed?"

"Because the North Vietnamese didn't use their fire-control radars until the last few seconds, sir. The Fan Song radar needs a minute to warm up, and then it needs a minute to acquire the target. The Radar Warning Receiver on the B-52 can detect the Fan Song, so the EW usually has a minute to acquire the signal, determine the frequency, and jam that frequency. From that point the Fan Song operator changes the frequency over and over again, and the EW has to adjust. So, it normally is a frantic exercise on both ends. But if the Fan Song only comes up after the missile is launched, well we've got a problem."

"How are they able to launch before bringing up the radar?"

"We don't know, general. Ernest has a theory that our Operations Staff disagrees with."

"Let's hear it."

"Ernest says opening the bomb-bay doors quadruples the Buff's radar return. He thinks that our procedures requiring the doors be opened sixty seconds early gives the bad guys an opportunity to find us using a normal, 'skin paint' radar. That way they can point the SAM in the general direction and fire up the Fan Song for any last-second adjustments."

"That leads to a few questions," General Ham snapped. "First, how does he know the radar return quadruples? Second, why do we open the doors sixty seconds early? And third, how do you point an SA-2 missile in any kind of general direction and expect to get a kill?"

"I asked him all three of those questions, general. And then I asked around the staff, just to see who says what. Ernest said the radar return quadruples because he asked our approach radar people and that's what they said. That is hardly scientific. So, I asked our operations staff and they said it has never been studied so they don't know."

"Hold on a second," General Ham said. "You mean to tell me we've been flying the Buff for close to twenty years and we've been dropping iron bombs since 1965 and nobody thought to check that? Is that what you are telling me?"

"Sort of, sir. The system program office in Ohio proposed the tests when the airplane was first certified and we told them no, since the aircraft was dropping nukes and everyone thought it was unnecessary."

"What a load of crap!" General Ham said, smacking his fist on his desk. "Go on."

"Yes, general. I remember the sixty second rule but never heard where it comes from. It comes from our Operations Staff. Nobody here remembers the rationale, but the speculation was it gave the Radar Navigator more time to troubleshoot the doors if they jam. I don't remember a single case of a D-model's doors jamming since I've flown it but there are folks who do."

"Well, you got to open the doors to drop the bombs," General Ham said. "Can we shorten the time?"

"Nobody I talked to objected to 30 seconds," Stephen said.

"Get the staff to buy in and publish the change," General Ham said.

"Yes, general. As for the third question," Stephen continued, "Ernest didn't know. But our Intel Division says the Soviets have developed an optical aiming system, but it still needs to get the SA-2 within ten or so degrees of the target. That's why they usually use two radar systems, one for azimuth and the other for elevation. Nobody here can fathom how that can work. You can't just point visually to a target at 35,000 feet moving 500 knots and hope to get close. Besides, Hanoi had a solid overcast last night. The staff is researching this."

"Anything else?"

"Yes, general. The only other thing my brother seemed really bothered by was the post target turn."

"What's wrong with it?"

"He said that always turning right makes us predictable."

"That's a good thing. Do you know about the first Buff lost on a combat mission?"

"Olive Zero Two?"

"No, it was seven years ago, but most of SAC has already forgotten. I was in TAC at the time, and we were shocked that the very first combat mission assigned to SAC resulted in losing two bombers before they got anywhere near the target. The tankers were late, and the lead cell decided to enter holding without telling anyone. They turned right into the next cell and two Buffs collided. Killed eight of the twelve crewmembers. Our biggest threat with so many airplanes in the sky is hitting each other. That's why we fly the same altitudes every mission. We make a right turn after bomb drop, every mission."

"Yes, sir."

"Anything else?"

"Yes, sir. This wasn't from my brother, but I've been hearing we are planning an op downtown."

"Well, the rumor is right, for once. But don't go confirming it. The President is afraid the new Congress is going to cut funding for the war when they show up in January. I think he's right. He wants something massive. I think his exact words were, 'it must be brutal.' Bigger than anything we've done before. We're calling it Linebacker II. The operations staff just started working on it yesterday."

"Yes, general. When I was at U-T, the theater commander tasked us to come up with a plan for Hanoi and Haiphong. The work has already been done."

"That's not their job!" the general said, the veins in his temples glowing red. "Dividing command authority is never a good idea. Especially in a war. Besides, the primary job of the B-52 is dropping nukes, not iron. If the North Vietnamese get their hands on a downed Buff, you can guarantee the Soviets will have it. And if they figure out the B-52's secrets, there goes our nuclear deterrence! The B-52 is a national asset, not a theater asset."

The general sat in thought when his secretary appeared at the door. "I've heard enough," he finally said. "Go downstairs and tell the Old Crows about the jamming problem and see what they can come up with. Dismissed."

* * *

"General Mayfield," Stephen said over the phone, "I need to spend some time with your most knowledgeable Electronic Warfare Officer with B-52 experience. General Ham has a few questions."

"So, you don't just need a wise old crow," the general said, "you need a wise old Buff crow."

"Yes, sir."

"Come on down, I'll have someone waiting for you at the door."

Stephen made the trip down five flights of stairs, into the basement, through an unmarked door. The cold and musty smell of the basement complex was replaced with a more sterile air-conditioned environment with no smell at all. He arrived at the vault entrance and was greeted by a lieutenant colonel.

"Don Crosby," the colonel said.

"Stephen Bellum, sir, thanks for your time."

"Anything for CINCSAC. Come on in."

While most directorates were fronted by ornate offices with double glass doors, secretaries, and professional paintings of airplanes or missiles, the Electronic Warfare Division had none of that. Their domain was an underground windowless vault lined by 24-inches of hardened concrete. Instead of a secretary they had an armed guard. Instead of a regal painting, their entryway had a drawing of a flight suit patch on a four-foot square piece of plywood. The painting was of a comic book crow holding an antenna and the logo: "The Crow Sez: Jam it!"

The room was crammed with desks, bookshelves, and electronic equipment. But there was no one there.

"Everybody out for lunch?" Stephen asked.

"I think when they heard you were on your way, they ran away like frightened mice."

"But not you, sir?"

"Oh, I would have scampered away too, but I was the one who received the call from General Mayfield. I had no choice."

"Well, it is nothing to worry about, sir. I just have a few questions generated from the crows at U-Tapao and Andersen."

"Like your brother?"

"Some, but not all. You can't blame Ernest for any of this."

"Of course. Fire away."

"Well, in order of the bomb run, the first question is why do we light up all the jammers at 125 miles?"

"Why not?"

"They say it gives the bad guys their first indication of where we are, giving them a warning about our timing and direction."

"Well, I guess that could be a problem, but you got to be pretty precise with a SAM, you just can't point it in the general direction and hope to get within acquisition range."

Stephen looked at him, expecting more. Crosby stared back, and then thought better of having a "no blink" contest with an officer who dealt directly with the four star himself. "Have them up channel it and we can do a formal study."

"Yes, sir. The next item is the post target turn. Always turning right makes us predictable and some say it could decrease the effectiveness of our jammers."

"Could? How do they know it could?"

"I don't know, sir."

"I've flown over a hundred missions and never had a problem. Maybe it is just combat nerves."

"They say the North Vietnamese have improved their tactics."

"Well bring us some evidence and we'll change ours." Crosby raised his voice slightly. Stephen thought he had struck a nerve.

It was easier going one-on-one with the lieutenant colonel than a lieutenant general.

"Next question, sir. Why aren't we allowed to use chaff?"

"Our studies say it is best saved for use against fighters."

"But the fighters aren't showing up at night, so why save it?"

"I suppose it can't hurt, but it takes a lot of manhours to reload. The electronic jamming is so much better, why even bother?"

"Yes, sir. One last question. How are the North Vietnamese able to launch with any probability of target acquisition without firing up the Fan Song acquisition radar at least a minute ahead of time?"

"They can't."

"We have three bomber crews who say that's exactly what they did last week."

"I haven't heard that. Let me look into it."

16. Conscientious Objector

Berkeley, California

"I thought your dad said the war is winding down!" Sherrie held the campus newspaper in her clenched fist. The paper included photos of bombed out villages, civilians fleeing burning buildings, and an especially menacing photo of the President. She slammed the paper down, causing her coffee to spill onto the table.

"He did," Martin said. "But he retired a few years ago so he doesn't really know. He sounded like he was pumped up on pain killers too." He leaned back in his chair, resigned to his powerlessness.

Sherrie opened the paper and flipped to the story behind the full-page cover photo of a bomb crater in the middle of a small Vietnamese village. As she scanned the text, Martin steeled himself for the next accusation. He was, he realized, shouldering the blame for events he had nothing to do with. The fact he opposed the bombings mattered little. *I am on trial for the crimes of others,* he thought

"Do you see what they are doing? They are bombing the north now! They are killing civilians!"

"I know, I know. Things have gotten worse, not better."

"Think of what your dad's accident did to your mom," she said, her voice softening. "And now she has had two sons over there fighting, and now it could be three. How much does one family have to give?"

"Stephen is stateside now. He might go back if the war drags on long enough. Ernest is in Thailand."

"You know what I am saying."

"I do, sorry."

Martin wiped the spilled coffee from the table with his paper napkin, allowing the moisture to soak through and cool his hands. The hot Berkeley summer had given into a cool California autumn. It was Martin's favorite time of year, but he had little reason to enjoy the atmosphere. *If I am the one on trial, my only option is to plea bargain.*

"Our family has given enough, I know that. But Ernest is over there of his own free will. Besides, my dad said the B-52 is practically invincible . . ."

"But," she started to interrupt.

"But they shot one down last month," he said, finishing the thought. "I know, I know."

Sherrie pushed her coffee cup aside and reached across the table to hold Martin's hands in hers. She spoke softly, looking into his eyes. "How can you be sure Ernest is there voluntarily? Maybe once he's seen what is going on, maybe he isn't so willing?"

"I don't know. It's next to impossible to call over there without some kind of super-secret military phone. But Stephen talked to him the day after. Stephen says Ernest is going to be okay."

"Are they close?"

"Not really. If anything, I'm closer to Ernest. We are all four years apart, so when I started high school, Ernest started college, and Stephen started with the Air Force. I hardly know Stephen, though I think I know Ernest. Stephen says Ernest is going to be okay."

"But can he guarantee it?"

"Nobody can."

"Which is why you can't let them take you," she said, softly. Her eyes began to tear up. *The defense throws itself on the mercy of the court,* he thought.

"Even if Ernest is going to be okay, what he is doing over there most definitely is not! Just because your dad was a war hero doesn't make war heroic." She held the paper for emphasis.

"You are right about that. My dad talks about his war as if it was a glorious saga, but I think he knows in his heart that by the time he got there, most of the fighting was done. That doesn't make him less of a hero, but it does set up his sons for following a false legacy."

Sherrie's shoulders dropped and she leaned back in her chair. "So don't follow in his footsteps," she said quietly.

"He thinks the war is going to be over soon," Martin said. "He might be right. But it is a chance I cannot take. I wrote my conscientious objector letter right after the election, but I never mailed it," he said. "I'll send it today."

17. The Lesson

U-Tapao, Thailand

"That's remarkable, Ernest," Marjorie said with a warm smile. "Are you sure you've never had a lesson?"

"Not with another human being," he said, holding the violin with his left hand and perched on his left shoulder. "I read a few books and I listen to a lot of music. I've even been to a concert or two. But I can tell it is far from remarkable."

She laughed. "Well, it isn't going to get you into Carnegie Hall, but the fact you could play anything at all with your posture and grip is remarkable. I think with a few basic techniques, you will be surprised at how good you can get."

"Well, I am your block of clay, mold me as you see fit."

"My first violin teacher called it the 'nose, scroll, and toes' posture. Let me show you." She took the instrument and nestled it between her shoulder and chin, holding the neck of the violin with her left hand. She was still in uniform and her long hair was neatly tied out of the way. "My nose points to the scroll and I can see my toes just over the violin. If my posture is correct, my head is level and so is the violin. Now you try it."

Ernest took the violin and held it under his chin firmly, gripped the neck even more firmly, and looked to her for approval.

"Relax," she said, circling behind him and placing her left hand on his, aiming the violin parallel to the floor. She placed her right arm under his to lift his chin slightly. "Like this. Perfect."

"I think I see," he said, hoping she couldn't detect his racing heart. "I guess I've been afraid I might drop it, so I put a death grip on it."

"No death grips! You cannot play with feeling if you can't feel your own fingertips. Now play that piece for me again."

Ernest looked at his sheet music and repeated his efforts, playing "Amazing Grace." He smiled at his efforts. "Better," he said.

"Much better," she said. "I am still impressed that you taught yourself to read music and got yourself to what I would expect from a second-year student."

After just fifteen minutes, Ernest had to stop. "My hands are cramping up. I haven't picked up the violin in a few weeks."

"Relaxing your grip will help, but you do need regular practice. Believe it or not, playing the violin is a physical workout."

"I believe it, how about playing for me?"

"I haven't played in a while, but it is only fair." She took the violin, made sure her posture was as she had preached, and played "Amazing Grace" as printed on the sheet music plus a few flourishes of her own. Ernest felt his eyes moisten.

"Beautiful," he said. "You play, what's the word I am looking for? You play emotionally."

"I'll take that," she said. "I think you bring that out in me."

"I don't know how to pay you for the lesson. Maybe I can give you some techniques with my soldering iron."

"How about you buy me dinner and we'll call it even?"

"Deal, in fact, it is about that time. Can I offer you the O'club or the O'club?"

"Hard to decide," she said. "You pick. But let me change first. Do you mind me in my civies even though you are in your flight suit?"

"No, not at all."

She left the room and he sat back and noticed for the first time her apartment was the opposite of his. They were both in the same complex of officers' quarters but different buildings. The furniture provided by the Thai Navy was comfortable, but basic. She had a small table with four chairs, a basic kitchenette, and a door to the bedroom and another door for the bathroom. Where his bedroom was on a wall right of the entrance, hers was on the left.

"All set," she said, popping out of the room.

"Wow," he said.

"Oh really? I approve that you approve." She wore a dress that came to her knees and shoes that were not quite sandals but not quite high heels. Whatever they were, they were not combat boots.

They left her room and walked along the concrete path from officers' quarters to the officers' club. "I've never been to the formal dining room," he said.

"I didn't know they had one, we can just do the bar."

"No, let's do this up right."

The formal dining room looked a lot like the informal dining room, except for the tablecloths and the fact the waiters wore ties. She chose a steak, a salad, and a Beer Singha and he doubled the order.

"Do you ever regret not pursuing music?" he asked.

"Maybe a little. I was a good student growing up and really fell in love with anything to do with science. But my parents pushed me into music, just as they had done with Caroline, my sister. Caroline loves it. I guess I wanted to do my own thing, so here I am. I gave lessons while I was in college for a little spending money, but nothing more than that. I am happy with my choices. Meteorology may or may not be my calling, but I am enjoying it for the moment."

"How did your parents handle you marching to your own drummer?"

"Surprisingly well. I thought they would be mortified. But in the end, they said, 'okay,' whatever makes you happy."

"I have a theory about that. Parents are so excited for their first child they invest all their hopes and dreams into having them follow their chosen path. By the time they get to the second child, they don't have the energy to fight."

She laughed. "Now that you mention it, I think you are right."

18. The Fuse

Hanoi, North Vietnam

•

"The facts are circumstantial, I am afraid," Captain Sergei LItvyak said. "It looks like we destroyed a bomber, but it got away. I have a map."

Nikolai sat at his desk as the captain unfolded the map showing Vietnam, Laos, Thailand, and parts of China. The paper had grown weak at its creases and Sergei carefully smoothed the area between Hanoi and the Laotian border.

"Just show me," Nikolai said, betraying his frustration.

"Yes, major. The attacks were here," he said, pointing, "at the railways in Vinh. Many missiles were launched and our allies say many bombers were hit."

"And yet we have no evidence of wreckage."

"Yes, but I think we can piece together at least one credible kill."

"Then where is the wreckage?" Nikolai said, betraying anger on top of his frustration. His subordinate recoiled into a position of attention, his spine erect, and his hands to his sides, as if reprimanded. Nikolai immediately regretted his outburst. "I am not angry with you, Sergei. Just tell me what we know for sure."

"Yes, major. It appears the bomber was heavily damaged and glided across the border with Laos." He drew a line with his finger from Vinh directly west, to just over the North Vietnamese border.

"And how do we know this?"

"We intercepted radio transmissions between the bombers in the same group, as well as from the damaged bomber to their helicopter rescue. It appears the entire crew bailed out successfully and the bomber crashed in Laos."

"A goose that is shot and lost is like a goose that has never been shot at all," Nikolai said, leaning back in his chair and closing his eyes. He was grateful that he no longer needed to wring the sweat from his towels only so he could sweat some more. But the ledger of good and bad was lopsided to the bad.

"I have never been a hunter, comrade major."

"Nor I," he said. "I grew up in the city. In Sofrino."

"Moscow?"

"Just outside. And now I am in this godforsaken neighborhood in this godforsaken country, seven thousand kilometers from home. There are twice as many of these bombers in America, all aimed at Mother Russia. And yet the goose gets away."

Sergei opened a notebook and pointed to a hand drawn triangle. "I spoke also with the missile battalion commander. I think he has a good theory about this goose."

"Taking credit, I suppose."

"Undoubtably, but also an idea about how to improve our odds of getting our hands on a future goose."

Nikolai moved forward in his seat to view the triangle, drawn with one side parallel to the bottom and a stick figure airplane above the top, vertical leg.

"At present the missiles are set to go off using a proximity fuse or when the fuel is expended, which ever happens first. The proximity must be about fifty meters of the target."

"Which is why the radar guidance is critical," Nikolai said. "So the only solution is to improve the radar guidance. But we must fight the enemy jamming to do that. That is why it is of such importance to get our hands on their countermeasures. Getting just one bomber will change the outcome of the war. Not just this war, but our war with the Americans. We must defeat the enemy noise."

"Yes, major, we are working on that. But our ally proposes something we can do as well to improve our killing effectiveness.

The missile can also be detonated after a certain time interval since launch."

"But that risks detonation too early, negating the proximity fuse. We have thought of this before."

"Yes, major. But the Americans have presented us with another gift." Sergei drew a line parallel with the bottom of the page and through the stick figure airplane. "They have a fondness for flying the same altitudes. They have done this with every B-52 attack since April."

"Exactly the same altitudes?"

"Between ten and eleven thousand meters, without fail."

"And what good does that do us?"

Sergei drew a two-sided arrow along the bottom leg of the triangle. "If we can approximate the distance over the ground of the bomber, and we know the altitude of the bomber, we can figure the distance from the missile launch point to the bomber itself. We can then know how much time it should take to be near the bomber."

"No, no," Nikolai said. "This is even less accurate than the proximity fuse."

"But comrade major, we are not shooting just one goose. We are shooting into a flock of geese. And if we shoot enough missiles into the flock, we will surely hit something."

19. Eve

U-Tapao, Thailand

"That was quite good," Ernest said as they left the base theater. "I've read a lot of his biographies and it looks like they got it mostly right."

Marjorie walked alongside Ernest, her right hand holding his left arm. The evening had turned cold, and the poorly lit street meant they couldn't walk at their normally brisk pace.

"Did he really slap the soldier as they showed it?"

"Oh yes. In fact, he did that more than just the one time. General Patton was known for giving people a quick kick in the ass and his verbal abuse was legendary. But he was the right man for the job at hand. Eisenhower actually sent him to Africa with instructions to kick the Seventh Army in the butt. And butt kicking was what Patton did best. But he didn't have much patience for soldiers he considered cowards."

"I wonder about the word 'coward' in the context of war, Ernest. How can you blame anyone for being scared when they are being shot at? You can hardly blame them."

"I suppose."

"Do you get scared, Ernest?"

"Too busy to be scared. Maybe that was part of Patton's magic. He kept his men so busy they were more scared of him than the enemy."

"The movie shows that he was flawed, but it still leaves me with the feeling he was a great man."

"A great man?" Ernest thought as he navigated the on again, off again sidewalk, careful to leave Marjorie with an obstruction and mud free path. She tended to walk with her eyes on him, leaving her safety in his hands.

"Yes, I think he was probably a great man," he continued. "When all this is over and we get back to the states, I have a few books about him. He was a man of contradictions. The movie left out a few important things about him."

"Such as?"

"He loved his family and wrote to them daily. I think he had a son in law who was a German POW. It is one of those ironies in life that some of Patton's words to the troops caused some of those troops to shoot surrendering enemies, rather than take them as prisoners."

"Really?"

"Patton said something to the effect, 'I cannot see the value of taking any prisoner alive.' The movie hints at that, but I think he was even more blatant about it."

"How awful."

"War is messy."

"Do you worry about it?"

"No."

"Well, I do, Ernest." She held his arm more tightly.

"Don't. My job is keeping us safe up there, and I know what I'm doing. Our gear is good."

They reached the officer's quarters complex, passing through an open gate in a wall surrounding what had become known on base as "officer's country."

"What's that smell?" Ernest asked as they passed the gate.

Majorie released her grip on his arm, knelt to the grass border, and found the top of a bulb. "It's wild ginger, a wonderfully fresh ginger."

"Green thumb?"

"Hardly, but I do like to cook. Maybe I'll whip something up for you in my kitchenette one of these days." She returned to him, reclaiming the arm she held since they left the theater. They continued walking.

"It's a date. But for tonight, I'm sorry we have to skip the violin lesson. I've got a long day tomorrow."

"Of course. Me too actually. I guess you boys are going to be busy tomorrow. We've got the entire weather shop on duty all day tomorrow. In fact, they've told us this is going to be going on for at least three days."

"That's what I hear too. In fact, let's stop by my room first. I want you to take my violin for a few days."

"No," she said, stopping. Why?"

"I won't have time to practice and having it nearby will distract me. Besides, you will make better use of it until all this is over. I have a good feeling that we are going to end this war in the next week."

20. Prelude

Bellevue, Nebraska

Jonathan sat in his favorite recliner, facing the fireplace which he lit with a combination of crumpled newspaper and lighter fluid. He had strategically selected the size of the logs from largest to smallest from the bottom to the top, except for a large log atop the square based pyramid. Stephen got up and moved his chair alongside his mother's, a few feet further away to avoid the direct heat.

"Is Ernest going to be okay?" she asked once everyone had settled their seating positions.

"He should be fine," Stephen said. "He's flying good equipment and he knows what he's doing. The D-model has the best ECM gear on the planet. The Viet Cong never laid a finger on me."

"That's good to hear, just when I could stop worrying about you, he goes there and now I have to worry about him."

"At least you don't have to worry about Martin."

"Martin? He's the one I worry about the most."

"Really? Why is that?"

"Well, with you it was pretty easy, honestly. You are like your father. You both have a clear idea of where you are and where you are headed. It may have been flying airplanes in wartime, but I always had the sense you both have an awareness about you. You collect good friends to help. You have firm convictions and aren't easily taken off course. That kind of thing." She gave Jonathan's arm a squeeze.

"And Ernest?"

"I never could figure him out," she said, smiling. "Sometimes he could make me so angry, with those word games of his. At first, I thought it was that he knew he was smarter than me and wanted to make sure I knew that. But that wasn't it."

"What was it then?"

"I still don't know. He's not one for friendships, is he? But at least I could rely on the fact that he has a firm view of his place in the world. He knows where he is, and he knows where he is going. Even if he keeps the rest of us guessing."

"I never thought of it that way," Stephen said. "So that leaves Martin."

"So that leaves Martin," she echoed. She sat back in her chair and looked to Jonathan, who appeared lost in his own thoughts. She picked up her cup of tea and stared into the swirling leaves. Stephen smiled at the thought of his mom researching tea leaves to unravel the mystery of her youngest son.

"Martin is very smart in his own way," she said at last. "But he is changeable. One day he thinks this, and the next he thinks the opposite. You built model airplanes as a boy, so he decides he likes to build model airplanes too. But then he loses interest and goes to something else. Ernest builds electronic things. So, Martin tries building electronic things but then loses interest in that. He reads one book and thinks he knows all there is to know about that. Then he reads another book, and all his opinions are changed. I always worried that he would fall into a bad crowd. I was hoping he could find good friends in college."

Stephen thought about the latest news footage of protests and sit-ins at college campuses around the country, but he kept those thoughts to himself.

"It's getting late, I think I'll leave you boys to discuss your secret thoughts," she said, getting up and pecking both men on their cheeks. Stephen and Jonathan sat quietly, and the fire finally started to take on a glow without the flames of a blast furnace.

"I guess SAC is getting serious about conventional warfare," Jonathan said, finally. "Never thought I would see the day."

Stephen considered that. Their discussions of what both considered "the family business" didn't stop when Jonathan retired, but became necessarily less specific when Stephen joined the headquarters staff.

"I think so," Stephen finally said. "But what makes you say so?"

"I still have my sources. It seems we have more than the usual number of bombers and tankers in theater. We've also ramped up support personnel. Looks like an early Christmas present for Uncle Ho."

"Ho Chi Minh is dead, dad."

Jonathan propped himself up in his chair to lean toward Stephen. "I know that. But President Nixon's political career will be dead if he doesn't end this war before January. It doesn't take a rocket scientist to add all this up." Jonathan lowered his voice, "All this with Ernest over there!"

"I know, I know." Stephen adjusted his chair a few feet closer to the burning fireplace and a few feet away from his father. He caught a view of the backyard. The snowfall was light, and the flakes were very small. "White Christmas this year?"

"That would be nice," Jonathan said, softly.

"If we did have a big operation going, and I'm not saying we do, but if we did, Ernest will be fine. The gear, the tactics, everything. The SAC Operations Staff has this thought out. Well, you should know that dad. You were on the Operations Staff for more than ten years. You know the people there."

"Yes," he said, looking into the fireplace. The largest log atop the pile suddenly split into two pieces, the largest of which held on for another second and then surrendered to the flames below. "That's why I'm worried."

21. Linebacker II, Night 1

U-Tapao, Thailand

"How can they expect us to get all this done when we just got the mission folder an hour ago?" Kevin asked. "I still don't have the target photos! How can I drop on targets I've never seen before?"

"Don't know," Major Stevens said. "That's the way they want it done this time."

Ernest sat in the back row of the crew bus quietly. It was all very odd. They usually got all their paperwork the day before the mission. For tonight's 2000 Time Over Target, they knew nothing about the mission until they showed up for the flight at 1500. Only Major Stevens got any advanced notice. He and the other aircraft commanders only knew an hour earlier. The bus turned into the command post parking lot.

"No, we're skipping this," Stevens said to the bus driver. "Head for weather and then the airplane." The driver stopped, circled the parking lot, and pulled back onto the main avenue.

"I can't bomb if I don't have photos," Kevin said.

"Don't worry about it," Stevens said. "We'll have your photos before we takeoff. But we can't do anything until we see Ernest's girlfriend first."

"Captain Bellum has a girlfriend?" Sergeant Polanshek asked. "Why doesn't anybody tell me these things?" Polanshek shot Ernest a thumbs up and a grin. Ernest felt a little embarrassed but couldn't help returning the grin.

Marjorie's weather briefing was thorough, but it was punctuated by her smiles to one member of the crew. The rest of the crew pretended not to notice, and Major Stevens thanked her on behalf of the crew. Ernest lingered as the crew filed out.

"I am starting to enjoy weather briefings for the first time in my flying career," he said.

"I am enjoying giving them more," she said. "Especially for one particular crew."

"See you tomorrow?"

"Of course."

Ernest was last to board the bus. The crew sat and read their folders as the driver wove them through the taxiways to their waiting bird.

This was Ernest's first trip "downtown" to Hanoi, but not his first trip north of the DMZ. He had spent hundreds of hours with the squadron's more senior EWs, some of whom had flown the first B-52 flights downtown back in 1967. Since most crews rotated every year and few came back for a second tour, only the crew's Aircraft Commander and Radar Navigator had been there before. Major Stevens was on his second tour and had well over 100 missions under his belt. Kevin was on his third tour and refused to discuss how many missions.

"Okay, listen up," Stevens said from the front seat. "We are flying number three in Brown Cell, which is the fifth cell in the first wave.

"First wave?" Harry asked. "How many waves are there?"

"Three," Stevens said. "We have a 2000 TOT. Wave two is at 2400 and wave three at 0400."

"They taught us at tactics school we should concentrate our forces when we can," Harry said. "Three waves, four hours apart. That's not very concentrated."

"We're talking a lot of airplanes," Stevens said. "For wave one, U-T is sending 48 bombers and I think Andersen is sending twice as many, maybe more. I bet the next two waves are at least that many."

"But why four hours apart?" Harry asked.

"I don't know," Major Stevens said again. "That's the way they want it done."

"This is the same route as the last time I went downtown," Kevin said. "Same altitudes, same routing, same turn points, same everything. This is the way every single trip downtown goes. We fly through Laos north towards China, then turn east into North Vietnam, then southeast down the Tam Dao mountain range to Hanoi. The only thing different is the bomb load."

"I was just looking at that," Harry said. "We are as heavy as we can be for these altitudes. Right above the stall. If we lose an engine before we drop our load, we are falling out of formation. Got that Ernest?"

"That means we lose mutual jamming," Ernest said. "So don't lose an engine, okay?"

"I don't get why we have to approach from the north," Tommy said.

"They said it takes advantage of the tailwind," Stevens said. "It gets us to the target more quickly and lessens our exposure to the SAM threat prior to the target."

"But it gives me less time to line up on the target," Kevin said.

"And it increases our time over the threat during and after the post target turn," Ernest said.

"I don't know," Major Stevens said. "That's the way they want it done."

"We only got eight chaff birds," Ernest said, looking at the ECM card in his mission folder. "Last time we had twenty-four."

"I think twenty-four is all they got," Major Stevens said. "Three waves mean we get one-third of the total." It still made no sense. The 24 F-4 Phantoms available for chaff duties would blanket the target area with hundreds of thousands of small strips of foil that would turn the enemy's radar into a big confusing sea of metal.

"I thought we had chaff," Tommy said.

"We do," Ernest said. "In fact, ours is better."

"You don't know that," Stevens said.

"Chaff from another airplane has to be timed perfectly to blanket the drop zone the same time we show up. If we used our own chaff dispenser, we know it is going to be on time."

"I don't know," Major Stevens said. "That's the way they want it done."

As the bus pulled up to the front of their aircraft, Stevens stood. Once stopped, Ernest remained seated, waiting for his five crewmates to file out of the bus and onto the bomber. With his helmet bag in one hand and a briefcase full of secrets in the other, Ernest trudged forward and off the bus. Reaching the aircraft entry ladder, he left his helmet bag on the ground and boarded with the briefcase. Bypassing the radar navigator and navigator level, he climbed to the top cockpit level and turned aft to his own compartment. There were no holes in any panels and no trash left behind from the previous EW. It was a good start. He returned to the ground to complete his external preflight and retrieve his helmet bag.

He found a broken antenna and flagged down the aircraft crew chief. "Already written up, captain. Check your 781."

"Terrific," Ernest said. He returned to the cockpit and forward to the pilots' seats where he found the plastic covered binder with the sterile name, "Air Force Technical Order 781."

Major Stevens looked aft from his seat. "Hey Ernest, did you see . . ."

"Reading it right now," Ernest said. "Down one jammer."

"Okay."

Ernest sat in his ejection seat and ran through the test sequence of his many receivers and jammers. Everything came up normally except the jammer connected to the broken antenna. He had eleven more, so they had more than the minimum for Ernest to report "ready" when Harry polled the crew. Only the radar navigator reported otherwise.

"All sorties, this is Charlie, start up and taxi to the runway and standby," a voice commanded on the SAC command post radio channel.

"Who is Charlie tonight?" Kevin asked.

"Colonel Stipp," Stevens said.

"The wing commander?"

"Himself. You know this is important if he has Charlie duties. Engine start checklist."

As the massive bomber taxied the two miles to their position in line, Ernest thought he had everything he needed, even without whatever it was they were waiting for. From his compartment he could sense the aircraft's lumbering turns and track their progress by listening to the crew interphone and the command post radio.

"All sorties," Charlie broadcast, "expect a vehicle with your bags. Have the hatch open and your nav downstairs waiting. Once you have the bags and you are ready to go, check in with me, call sign only."

Ernest's ears popped as Tommy opened the entry hatch. "Here he comes," Stevens said. A moment later Tommy reported, "we got it, sir. Hatch secure."

"Brown 3 ready," Stevens reported.

"Mission is released, happy hunting," Charlie said after the last airplane reported ready. Ernest started counting the number of times the earth shook from the thrust of eight engines coming to full "dry" thrust, the boom of the water injection systems increasing the power to "wet" thrust and the noise to a level no man had ever contemplated. He knew an aircraft was rolling by every 90 seconds. It was not unlike a symphony, with Charlie conducting from his staff car at the end of the runway. But that only lasted five minutes.

"Rose two aborting, no water injection."

"Rose 2, Charlie, hurry it off the runway. Rose 3 you are now Rose 2. Brown 1, you are now Rose 3. Everybody moves up a spot. Call your position as you cross the threshold, not before."

"Brown 2, ready," Stevens reported as they crossed the runway threshold.

"Released," Charlie said. As the power came up, Ernest checked his clock and recorded the time into the crew log. They were eighteen minutes late. "So much for chaff support," Ernest said without keying his interphone button.

Without an attitude indicator or altimeter, Ernest could only track their climb with the seat of his pants and the chatter on the interphone and the radio. He could sense the right turn, knowing they took off to the south and would have to turn right. But he always tried to visualize the join up with the rest of the formation while still in the clouds, it was a chance for bad things to happen.

"Gunner, pilot," Stevens said, "any sign of Brown 3?"

"I'm Popeye," Sergeant Polanshek said.

"Popeye?" Ernest wondered. He knew the gunner sat under a glass canopy in an extension of the tail, aft of the aircraft's vertical fin. He had a magnificent view of everything behind the aircraft and could also see both wings and most of the eight engines. But he would have to rely on his guns' radar to see while they were still in the clouds. "Oh," Ernest said to himself, smiling. "He's looking at the world with one eye shut. Clever." The buffeting stopped and the ride smoothed.

"Pilot, gunner. I've got him now. Five, six miles left."

"On top at 10,000," Harry said, "just like the lady said." Ernest leaned left from his seat and caught a glimpse of the full moon shining through the pilots' windows. "Ernest, you ought to marry that gal."

"She's cute," Polanshek said. "For an officer."

"Hey Ernest," Tommy said. "She got any sisters?"

"Yeah, but she's too old for you," Ernest said.

"Okay, let's focus on what we need to focus on," Stevens said.

"You will never guess what our target is," Kevin said. "It's Radio Hanoi. We are dropping 60,000 pounds of bombs on a radio station."

"Well, I guess they got their reasons," Stevens said.

"We got to be careful," Tommy said. "There are POW camps around there. And we also got a 150-degree post target turn. The weather girl said that turn is going to be into the wind. And I just got to the page with the active SAM sites. Fifty of them."

"You copy that, E-dub?" Stevens asked.

"Roger," Ernest said. "I have the same package. Maybe we should do some of that pilot magic as soon as we drop."

"You heard the same thing I did during the briefing," Stevens said with a hint of irritation. "No evasive maneuvering for thirty seconds. We fly the plan."

"Pilot, gunner. Take a look, to your right."

"The rotating beacons on the Andersen birds," Stevens said. "They are sending twice as many as we are, but they have to cross over South Vietnam to get into line behind us."

"So, a 12-hour mission with air refueling gets even longer," Tommy said.

"I can see rotating beacons in front of us too," Harry said. "All the way to Hanoi."

"We are about to turn south," Stevens said. "Let's knock off the chatter."

Ernest rechecked his AN/ALR-20 scope, an 8" by 12" screen that showed all North Vietnamese electronic signals, their strength indicated by the height of a strobe. There was definitely activity. A second scope, an APR-25, showed the range of each signal. The art to his job was about to begin, watching for the pattern to change and then finding the frequency to jam.

"IP in ten minutes," Kevin said. Ernest rechecked his scopes. There was so much jamming going on it was difficult to tell where each signal came from. In theory, the jamming from each aircraft in the cell would combine and make it that much harder for the SA-2 to "burn through" the noise and home in on any of the bombers.

"The ground is lighting up ahead," Harry said.

"SAMs?" Ernest asked.

"No. More like bombs ripping a city in two. There's more. Yeah, no doubt about it. Lead cell is over Hanoi."

A swishing noise filled Ernest's ears. The audio signal was verified by a light on his RWR. "Triple A," he announced to the crew. "We are too high for them, but that doesn't stop them from trying." Anti-Aircraft Artillery emplacements used radar to target the bombers, the AN/APR-25 Radar Warning Receiver detected it a millisecond before Ernest's ears made the same determination.

"IP," Kevin said.

"Too many knobs," Ernest often sung to himself when in training. The ALT-22 was the best they had, but it was a manual jammer. It took three knobs to get the right frequency and that was . . . too many. This was no longer a game. He couldn't detect what was enemy radar over his own jamming signals, but the rules said jam everything.

"Ninety seconds," Kevin said. Ernest looked in vain for the telltale Fan Song radar. There was so much jamming from each aircraft in the cell, it was difficult to see anything. But the twin strobes of the Fan Song would be hard to miss.

"Sixty seconds, doors," Kevin said. The aircraft vibrated with the turbulence of the huge slabs of aluminum thrown into the wind. Other than the vibration, the airplane felt still, the pilots taking care not to disturb the aircraft's altitude or heading to the target.

"Thirty seconds," Kevin said.

Ernest closed his eyes briefly, hoping to divert the energy to his ears. It worked; he heard the rattle. "Fan Song," he said, spotting the twin strobes piercing his scope, looking like the fangs of a snake to accompany the sound. Keeping his eyes on the scope, he reached left and up to his first jammer and tuned in the frequency. Then he assigned the same frequency to another jammer, and another.

"Twenty seconds. Stop maneuver. Level," Kevin said.

"Come on, come on," Ernest said to himself. The frequency changed and Ernest attacked it with another jammer.

"Bombs away," Kevin said. "100, 90, 80, . . ." he counted. The aircraft shuddered as 108 bombs began their journey to the surface.

A buzz in Ernest's ears got his attention. And then another. He pulled his head out of the scope in time to see the RWR light. "Launch," Ernest yelled. "We got a SAM heading right at us."

"Looking," the copilot and gunner called out simultaneously.

The fangs reached the first concentric ring on the scope. "Uplink," Ernest said. "The missile has guidance."

". . . 20, 10, release complete, standby," Kevin said.

"I got one," Harry said. "Make it three, coming out of the clouds right now."

"Heading right at us," Sergeant Polanshek said.

The fangs were engulfed in a sea of noise. Ernest watched as the fangs reached the third ring, still pointed right at them. "Three rings, SAM has locked on us." He moved his hand to the next row of panels and grasped the first chaff dispenser button. *Do it,* he said to himself, *it can only help.* But he left the chaff dispensers idle.

"Thirty seconds, maneuver," Kevin said.

Ernest started to speak but the G-forces from the 60-degree bank turn pushed his oxygen mask off his face and he struggled to lift a hand to re-cinch the mask. The fangs were engulfed in a cloud of noise, rising and falling. *Will you just die?* Ernest said, switching frequencies again. The G-forces returned to normal as Stevens rolled the wings level. Ernest added his last jammer to the equation and reassigned his first jammer to the new frequency. The Fan Song signal finally lost the battle.

"Missiles diverging," Sergeant Polanshek said. "Headed where we used to be. Living the good life back here, gentlemen."

A new sound caught Ernest's attention, one he had never heard before, except in training. He checked his scopes to be sure. "Fighters," he said.

"Where?" Polanshek asked.

The signal on the scope disappeared but the sound persisted.

"I don't see anything," Polanshek said, frantically. "Where are they, EW?"

The signal was faint. "Seven o'clock," Ernest finally said. "It looks like two of them, ten miles, closing."

"Got 'em now on radar," Polanshek said. "Off our left, heading for Three."

"Brown Three, Brown Two, check your seven," Stevens said on their interplane frequency.

"Cheezus!" Brown Three said.

"Closing," Polanshek said. Ernest watched the signals grow and added jammers to those he knew were already coming from Brown Three.

"I don't see nothing," Polanshek said. "No, I see muzzle flashes."

"Whose?" Steven asked.

"The sweet muzzle flashes of fifty cals," Polanshek said.

"Got him," Brown 03 said. "The other one is bugging out. We owe you a beer, Two."

Ernest scanned his sensors over and over again, finding enemy radar signatures and assigning jammers to them, but with each find there was no doubt the signals were lessening. Even the air-to-air signals had died out. They were leaving downtown. Unscathed.

22. A World Away

Omaha, Nebraska

"Let's go!" General Ham said from the apex of the command center, sitting in the seat normally reserved for a lower tier general, but now dedicated to the man himself. He was flanked by the operations staff. Stephen was seated in the back row of the outer ring but had a clear view of most of the generals at the table. Everyone was in their "Class A" uniforms, per CINCSAC orders. The war was to be a formal affair. It looked like Ham had his cheerleader face on, ready to rally the troops with a rousing speech. All he needed were some stats.

"Second wave is on target right now," Lieutenant General Stringfellow said. "Initial reports from the first wave are coming up on screen." An orderly placed an acetate slide on a projector and the room fell silent.

"This is a preliminary report on the first wave, and we still need to verify the details," General Stringfellow said. "But so far it looks like four B-52s were hit, one was downed, two are missing, and one of our Guam birds made it to U-Tapao and everyone survived."

General Ham stared at the slide. Nobody dared to speak. "Tell me what we know for sure."

"Reports are that SAM activity is extremely high," General Stringfellow said. "Hundreds of SAMs. Lilac Zero Three was hit seconds prior to bomb release and the bomb release jammed. The pilot was blinded by shrapnel and the copilot got hit in both arms. But between the two of them, they got the airplane out of there and landed it at U-Tapao. They say they stopped counting the holes in the airplane at 680."

"The second aircraft is a confirmed loss," he continued. "Charcoal Zero One was hit during the bomb run and exploded. The second aircraft in the cell did not see any chutes. Two other birds are unaccounted for, but they are still airborne for the long flight back to Guam, so it is too soon to count them out."

"Find out what happened," General Ham ordered. "Did our crews fly the plan? Did the ECM fail? What? Get me answers." Ham fell back in his chair. "Was the moon a factor?" he asked.

"Sir?" General Stringfellow asked.

"The moon!" General Ham said. "The weather folks said it was going to be a full moon last night and I asked them when the moon was coming up over Hanoi and when it would set. And I am still waiting for that answer. So, I want to know, was the moon a factor in getting our airplanes hit?"

"Ah, I don't know," General Stringfellow said. "I'll find out. But we would like to address tomorrow night's plan so we can get the folders out to 8th Air Force in time for Night Two."

"Yes, yes," General Ham said. "But let's settle this moon question first."

Stephen, in the back of the room, decided his time was better spent on the phone to Thailand. He returned to the command center four hours later just as the next session began and sat next to Colonel Snedeker in the back row.

"Better or worse?" Stephen whispered.

"Both," Snedeker answered. "Peach Zero Two was hit by a SAM during their post target turn, setting two engines on fire. The crew managed to make it to Thailand and ejected seconds before the aircraft consumed itself in flames. Everyone was picked up by helicopters."

General Ham entered and the staff came to attention. "Be seated," he said.

"We are getting reports of G-model bombing problems," General Stringfellow began. "Almost half the bombs are coming back because the bomb releases are malfunctioning, or the crews are unable to get on course after turning to the targets because of the high crosswinds."

"We can fix the bomb releases," General Ham said. "Tell me about the crosswinds."

"The radar navigators are saying they don't have the time needed to correct to course when they make the turn south to their targets," General Stringfellow said. "ROE says if you aren't on course, hold the bombs."

"Yes, yes. But how is it that most of the crews are getting on course and the others aren't? We haven't had any reports of collateral damage. So, what gives?"

"The 8th Air Force Commander suspects crews are making evasive turns prior to bomb release," General Stringfellow answered. "That is strictly against our instructions. He's going to crack heads on this. But there is something else, on the collateral damage front."

"What?"

"The North Vietnamese say one of our bombers made a direct hit on the international airport. Took out their runways and leveled the civilian terminal. The Soviets say two of their airliners parked there overnight were destroyed."

"Keep me posted," General Ham said, rising. "I am expecting better news in four hours."

As he got up and he made eye contact with Colonel Snedeker. "You two," he said, "in my office now."

General Ham sat at his desk and set his eyes on his two subordinates, standing at attention. He looked directly at Major Bellum, ignoring the colonel. "You had any words with the troops?"

"Yes, general," Stephen said. "It isn't pretty." General Ham sat, motionless, so Stephen continued. "The pilots are incensed that we are flying the exact same altitudes, routes, and turn points into Hanoi as we always have, and we are doing that cell after cell after cell. They say we are making it easy on the SAM sites to pick us out of the sky."

"The operations staff say we got to do that to keep our airplanes from hitting each other," General Ham said.

"We tested an approach called 'basket weave' two years ago," Stephen said. "We hit the target from multiple directions, like reeds in a basket. We keep aircraft separated with timing and altitude. 8th Air Force proposed this again two weeks ago, but the operations staff rejected it."

"Well, we can't go changing the plan at the last minute," General Ham said. "What does your brother have to say?"

"I think Ernest is as confused as everyone else by the failure of the ECM," Stephen said.

"Whoa," General Ham protested, raising both hands. "Don't say the ECM has failed. We lost three percent of our first two waves, that means it worked at least ninety-seven percent of the time."

"Ernest said the time it didn't work for them was just as they started the bomb run," Stephen said. "He says that for the most part, the gear works like it is supposed to. But it is failing at the worst possible time, enough to cost us airplanes."

"Explain to me how we expect the SA-2 to work," General Ham said. "Keep in mind I'm a fighter pilot. Give it to me at that level. I am sick and tired of the computer science talk I get from the Electronic Warfare Division."

"It all starts with locating the target," Stephen said. "The first radar is called the P-12 Spoon Rest, it is a simple search radar. It can see out about 300 miles and the operator calls the SA-2 site with the position of the target. It tells them where to point the fire control radar. That's the radar we call Fan Song. The Fan Song has a range of about 70 miles, which is more than the missile, so it is good enough. Once alerted, the Fan Song goes into search mode and it uses two trough-shaped antennas mounted at 90-degree angles to get both a vertical and a horizontal beam. The beams are shaped like fans, hence the name. Once it has the target, a smaller, narrow beam antenna transmits data to the missile, guiding it."

"So, what has gone wrong with the way it is supposed to work?"

"We aren't always seeing the Fan Song. Our EW's first indication of a SAM launch is quite often the uplink signal from the SA-2 site to the missile. That only starts six seconds after launch, because the uplink antenna is covered until the first stage is jettisoned. So, our EWs have reduced time to find the uplink frequency and jam it. The SA-2 operator then changes the frequency over and over again. It seems they have been able to burn through the jamming at times."

"But not all the time," General Ham said.

"Yes, general, not all the time. So that is actually the second question, why can they sometimes but not always burn through our jamming? The question that comes before that is how can they launch the missile in the first place without first activating the Fan Song?"

"And?"

"We don't know. The Electronic Warfare Division is studying it. My brother seems genuinely confused by it. I don't think anything has ever confused him before."

"How is morale?"

"Pretty low."

"Staff said it's high," General Ham said.

"I think the staff is wrong," Stephen said. From his periphery he saw Colonel Snedeker turn to him with a glare. Stephen had clearly crossed a line he shouldn't have, but the line was crossed. "Ernest said their target was a radio station. He isn't even sure they got it."

"Well let's hope the third wave goes better than the first two," General Ham said. "Dismissed."

That evening Stephen returned to the command center before the next meeting was set to start. A grease pencil scorecard was erected opposite CINCSAC's seat. An orderly was updating the numbers: three B-52s shot down, two heavily damaged. Rose 01 was added to the list.

"Another loss," Stephen said to himself.

"Three of the five were hit just prior to bomb release," General Ferrell said while entering the room behind Stephen. "The other two during their post target turns."

"Is that causation or proximity, general?" Stephen asked.

"Too soon to tell. The majority of SAM sites are near the targets and that is where the bomb release and post target turns happen. But I can see how a pilot would jump to the causation conclusion."

23. A Kill

Hanoi, North Vietnam

"Are you sure?" Major Nikolai Yakolev asked.

"Yes major," Captain Sergei LItvyak answered. "It was about 2030 last night, over the Yen Vien railroad yards. The 59th battalion used our triangulation method precisely, and we believe we got the B-52 aft of the cockpit, forward of the wing. We opened the aircraft up like an egg."

"How do you know that?"

"The cockpit fell to earth almost intact. The rest of the airplane rained on the area in thousands of pieces."

"So, we have the cockpit? Including the counter measures suite?"

"I have not visited the wreckage as of yet, major. But I am told the aircraft forward of the wings has not been destroyed by the fire."

"It is crucial that we keep it that way, Sergei. This will be our first look at their jamming equipment."

"Their jamming equipment was not so effective, in this case."

"That is true. But even if we can confirm the two other kills, that means we got three and failed to get a hundred more. If we can learn from the gear in this cockpit, perhaps. . ."

"Perhaps we can reverse the ratio of killed to not killed," Sergei interrupted.

"Yes, precisely."

Nikolai brushed the dust from his desk and clapped it off his hands. "It was intense, last night."

"Yes major, the explosions felt as if they were next door to my quarters."

"Were they?"

"The closest impact was a kilometer away from me, but I had never experienced such hell in one night."

"And then it began again," Nikolai reflected. He thought back to his own night of hell, feeling the earth shake and the walls of his temporary quarters flexing from the shock waves that seemed to envelope the city. "I thought after the first attack that surely they must have sent everything and began to relax. I did not sleep, but I did not expect it to begin again four hours later."

"And four hours later again."

The two men sat, silently, as if reflecting on a shared post traumatic stress. They were as comrades in foxholes, but the incoming bullets were massive bombs from the heavens.

"What of their targets?"

"Their accuracy was not very accurate. Except for Gia Lam."

"The civilian airport?"

"Yes, Major. They destroyed two of our parked airliners. No news of fatalities, however. Perhaps we can try the aircrew for this."

"The aircraft with the cockpit we have now, what of the crew?"

"All of the ejection seats have been used. We have captured at least three of the crew."

"Bring them to me."

24. Unnoticed

Berkeley, California

Sherrie's eyes were red from her overnight cry. She clutched the campus paper, her eyes imploring him for an answer. She had ordered her coffee with two sugars out of habit, but it sat untouched in front of her.

"They say Nixon is trying to end the war before Christmas," Martin said finally. "He wants to bring the POWs home before the new Congress meets. One of the big papers did a poll of the members of Congress. They say it is a done deal, they will cut funding for the war when they come back in January."

"I don't know what to believe. Did you see this?" she asked, pointing to the photo of a bombed-out airline terminal.

"I don't know. It is hard to believe."

"Whoever did this is a war criminal! The American Air Force is guilty of killing innocent civilians! This is genocide!" She gasped for air.

"I know, I know. The weird thing is that our campus anti-war group isn't meeting today. There are no protests anywhere in the country. Maybe the public thinks this is a good thing. A way to end the war sooner."

"Or maybe the stupid American public is more interested in Christmas shopping than their government turning an innocent third world nation into a fourth world hell." She got up, and without giving Martin another look, stomped out. He was considering leaving until he spotted Gene Phillips enter just as Sherrie left. Gene spotted him and walked over and took the seat she vacated.

"Sherrie doesn't look too happy," he said.

"The war."

"There is a lot to cry about, I guess." He caught the attention of a waitress and ordered a coffee. "But not for me. I just had my CO interview today."

"And?"

"I think it went very well. Much easier than when my brother did it. In fact, they told me that I was a model conscientious objector. I should have a new draft card with CO status in a few weeks."

"That's good. What were the questions like?"

"Just what I had expected. But the one I didn't expect was the last one. They asked, is there anything I would think is worth dying for."

"That's a tough one. Suppose your mother was at gunpoint and the offer is you or her?"

"I didn't think of that," Gene said. "That is a tough one."

"So did they give you any hypotheticals? What did you say, anyway?"

"I said no, there is nothing I think is worth another human life. They thanked me for my time, told me I would be approved for CO status, and they sent me on my way!"

"That's good. Congratulations. I put my letter in."

"Good luck, I think you are going to need it."

"Why do you say that? I thought you said it was easier."

"Easier for me, because I have the right mindset," he said, pointing to his temple. "I believe war is immoral."

"So do I," Martin protested.

"That's not what Sherrie told me last week. She says you think some wars are bad, and some wars are good."

"That's not what I said."

"Doesn't matter, it's what she thinks you said. You give off that vibe. They will see right through you."

As Martin got up, Gene said one more thing. "I've seen that look before, the look on Sherrie's face. I think she is done with you."

Martin walked to his next class from the café with his eyes on the ground, hardly noticing the world go by until he ran into another pedestrian. “Hey, watch it!” the person said without slowing down. He picked his eyes up and for the first time noticed the street was adorned with Christmas ornaments and filled with shoppers in a festive mood. The stupid American public.

25. Linebacker II, Night 2

U-Tapao, Thailand

It was a foreign language, this EW thing. It was no different than any other foreign language. Since living in Germany as a child, Ernest's thought pattern could switch from English to German instantly on a subconscious level. When speaking in German, his thoughts were in German. When listening to tourists or novice speakers he had to school himself to be patient, knowing that for most people speaking a non-native language, there was an extra step involved: hear the words, translate, think. So it was with the EW language at first: see and hear the signals, translate to frequencies, understand the implication, translate to actions with jammers, flares, or chaff.

"What frequency is that?" the instructor would ask.

"Two thousand, three-hundred Megahertz," the answer could have been. "G band," he could say for extra credit.

"What does that mean?"

"Fan Song."

"How do you counter that?"

The schooling drove that kind of methodology and it was tiresome. Ernest learned to see and hear and instantly move his hands to the correct jammer and act. It saved the question-and-answer drill and connected his hands to his brain. It was easy and it amazed his instructors.

But it didn't work like that in the real world. In the simulated world, the signals came iteratively, one by one. They happened repeatedly, but at a rate where you could isolate each step of the process. Once you developed some proficiency, they would come faster and if you were *really* good, they would combine them. But it wasn't like that over Hanoi. There were just too many. Too many signals. Too many sources. Too many.

Ernest bolted upright. There was a knocking at the door. What time was it? His flight suit was on the floor next to the bed, where he left it. Beyond that was his scarf, his boots, and his hat. He padded to the door and opened it. It was Major Stevens.

"It's a 1900 go," Stevens said. "Same song, second wave this time around."

"What time is it now?"

"You have a few hours. Report to the briefing at 1600."

* * *

Seated in the middle of the large mission briefing room, Ernest had to turn in his seat to see that there were half as many crews for night two. Most everyone present were trading stories of night one. Ernest made eye contact with a few of the other EWs but failed to sense any meaning from the fact everyone had something to say except the EWs.

"Room, atten-hut!" Everyone stood and caged their eyes forward. Colonel Stipp climbed the few steps to the stage.

"As you were," he said. "Yesterday we lost three B-52s. They were all from Andersen, two Gs and a D. We took some damage here, but all our birds made it home. We have a distinct advantage here since we are less than three hours from Hanoi where the Guam birds have to fly all day just to get in line behind us. Stick to your procedures, stay in cell position so you have good mutual jamming, and make sure those bombs are headed for their targets and no place else. We just need to keep this up a few more days and maybe we can get our POWs home in time for Christmas." The colonel paced the stage, surveying the crews. From Ernest's perspective, near the back, everyone was paying very close attention. Usually at least a few would be looking down, taking notes, or perhaps daydreaming. But not now.

"I am afraid we are still waiting for our mission packages," Stipp continued, "but we have been assured they will be here in time for your launch. One bit of good news. We've just confirmed that Tech Sergeant Turner flying on Brown Zero Three last night bagged himself a Mig!" The audience erupted in applause and chatter. All eyes searched for the gunner. "He's not here, he'll be on the third wave. But when they came off target a pair of Migs set up chase. Sergeant Turner blew the first one out the sky and the second one scrambled on home with his tail between his legs. So gunners, keep your scans up. The Migs are out there." Once the commotion ebbed, he finished, "any questions for me?"

"Don't even think about it," Major Stevens whispered to Ernest.

"Fine," the wing commander said. "I'll leave you to your mission prep." He departed the stage and everyone stood again. The mission briefer took his place.

"As you were," Lieutenant Colonel Heller said. "First slide."

The crews sat in stunned silence. "The same route," someone said. "Same TOT," another blurted out. "Same everything," Ernest said.

"Not exactly the same," Heller said. "We have half as many going tonight. The first wave is an all Andersen show. We pick up the heavy lifting for the second and third waves."

Slide after slide Ernest noticed the repetition and started to wonder why the crews had fell silent. No anger? No outrage? What was it he was sensing?

"Some good news," Heller concluded. "You might have heard that the message from on high was no evasive action would be permitted prior to bomb release. That word did go out to the first wave, but it was a mistake. You are allowed to evade missiles so long as you are flying straight and level for the last twenty seconds prior to bomb release. Questions?"

"Chaff?" someone called from the back row.

"No chaff. Any other questions?"

He paused. The crews were silent. "Godspeed."

As they filed out of the room, Ernest studied the look on the faces of his fellow EWs. “We fight them until we can’t,” one said. Ernest thought of the word he was looking for. Resignation.

“This is nuts,” Harry said once the crew was alone in the bus. “We were sitting ducks up there, flying right behind the next cell so all the SAMs had to do was shoot at the same spot, over and over again. Sooner or later they just get lucky.”

“Their good luck is our bad luck,” Tommy said.

“Exactly,” Harry said. “Now this is . . . well I don’t know what it is. But I don’t like it. How can the wing king let this happen? I thought he was one of us.”

“He is,” Major Stevens said. “I heard from one of his executive officers that he was practically yelling into the phone, trying to get SAC to change tactics. But no dice.”

“We did okay,” Kevin said. “Notice how the Gs had twice as many losses even though we sent four times as many Ds? The ECM gear matters.”

“We are definitely in better shape than the Gs,” Ernest said. “But it was a real mess last night. I’m not sure how effective I really was.”

Kevin looked at Ernest and smiled. “We’re here, aren’t we?”

* * *

The weather briefing was a replay of the day before, but Marjorie’s ever-present smile was noticeably dimmed. As the crew filed out, she rushed from behind the counter and grabbed Ernest by an arm and pulled him close, hugging him. “I was so worried when I didn’t see you this morning.”

“We didn’t get in until late,” he said. “I didn’t want to wake you and then I overslept.”

“Put a note on my door then, something, let me know you are okay,” she said, reaching up to kiss him.

“You two!” a voice from the next crew said. “This is a military operation, isn’t it?”

Ernest looked up to see the wing commander. "Yes, sir." He looked at Marjorie and smiled. "Don't worry. I'll put a note on your door tonight."

"Nice going, Ernest," Kevin said as he boarded the bus.

"What's Colonel Stipp doing at weather?" Ernest asked.

"He's flying on the next cell," Stevens said. "Sitting on the jump seat."

"He's flying over Hanoi without an ejection seat?" Ernest asked.

"I told you before," Stevens said, "he's one helluva wing king."

* * *

As the crew bus pulled up to the airplane, Harry noticed it first. "Every mechanic on base is here." Where each bomber was usually assigned three mechanics, there were six, each polishing various parts of the airplane, inspecting areas with flashlights, or standing in front with the aircraft forms, ready for the crew.

"They want to be a part of something big," Stevens said. "They are showing us that they will be with us in spirit."

Ernest went through his normal preflight routine, including leaving his helmet bag on the ground near the crew entry chute. After he finished his exterior preflight, he looked for his helmet bag. It was gone.

"Go get 'em, sir!" The crew chief held his helmet bag for him and saluted.

Ernest returned the salute. "Thanks, chief."

Go get 'em, Ernest thought. Go get who? Their target for the night would be, once again, Hanoi Radio. Apparently, they had missed on night one. The preflight salute was normal for the pilot. This was a first for Ernest.

"Ivory two, ready," Major Stevens answered when the "all call" came from Charlie.

"All aircraft," Charlie said on command post frequency, "slip all times twenty minutes, two zero minutes. Acknowledge." One by one each aircraft stated their call signs, including Major Stevens.

"Twenty minutes?" Tommy said on the crew interphone. "What's that all about?"

"I bet Andersen ran into some headwinds," Stevens said. "They got the first seven cells this wave. We have to fall in line behind them."

"Late. Just like yesterday," Harry said.

"No chaff support," Ernest said.

"Let's stop bickering over what we can't control," Stevens said. "Keep your focus."

The takeoff to the south, the turn to the north, the solid cloud deck at 10,000 feet, and the full moon. It was all the same, Ernest thought while staring at his radar receivers.

"Rotating beacons ahead," Harry said. "You can relax, Tommy. We'll just follow the red brick road."

"This is me, taking the night off," Tommy said.

"IP in ten minutes," Kevin said.

"EW, pilot," Major Stevens said. "You are awfully quiet. What are you seeing?"

"It's just like last night," Ernest said, "only less of it. I got five or six signals, jamming them. But it's not bad at all."

"Switching to the strike frequency," Harry said.

"It's okay, it's okay," they heard on frequency. "The SAM self-detonated and we took some shrapnel, but with a little sheet metal work we'll be as good as new."

"No losses!" another voice announced.

"Radio discipline," a third voice commanded.

Ernest permitted himself a sigh of relief. His scopes were lit up with enemy search radars and the jamming from his aircraft and that of his cellmates, but there were no Fan Song signals to be seen.

"Here come the doors," Kevin said.

Ernest closed his eyes and listened for his heartbeat. It was his normal method to estimate the time, but there was too much noise in his headset. He heard the gunner's radars ahead and behind. He heard the search radars. But still no acquisition radars. No Fan Song.

"Twenty seconds, level wings," Kevin said.

The pitch from the jammers in the lead aircraft lowered an octave and then went silent. Ernest opened his eyes to confirm that with the scope but there was too much noise.

Kevin keyed his interphone microphone. "Five, four, three, two, one, bombs away!" The aircraft jumped in rhythm with Kevin's call of the emptying bomb racks. "100, 90, 80 . . ."

"Fan Song!" Ernest called. "Launch!" The fangs on his scope towered over the three rings; the rattle was unmistakable. "I got two, three uplink signals, jamming."

"Got a single missile in sight," Harry said.

"Check again," Sergeant Polanshek said. "I've got two of them."

"That's what I got," Ernest said. "Jamming."

"All bombs gone, doors closed, clear to turn!" Kevin said, ignoring the thirty-second rule.

"Turn, turn, turn!" Ernest yelled as he felt the G-forces increase.

"Missiles going inside of us," Harry said. "Ease up on the bank, major. Missiles are headed for lead. Man, he's got a lot of bank in there."

Ernest kept his ALT-28s on the missiles' uplink frequency, hoping to rob them of the ability to guide on the beam sent from the Fan Song radar. But at least one of the missiles continued to guide. Ernest reached for every jammer he had left.

"Lead's hit," Harry said. "He's got an engine on fire and part of the tail is missing. There is fuel pouring out of the wing."

"Heading," Tommy said.

"Rolling out," Major Stevens said. "Ivory One is dropping back, we are lead now. Heading for home."

The ride home was quiet. Ernest finished his paperwork and stared at his six Hallicrafters AN/ALT-28 jamming transmitters. The ALT-28 was the best in the business, better than anything the Soviets had. But it was still a manual system. He needed just a second to determine the frequencies and a second more to tune each ALT-28. But his threat receivers seemed to be blind until the last few seconds, trimming the margin to zero. To make matters worse, the jammers worked until they didn't, and then they worked again. It wasn't like this in training.

"Ernest," Major Stevens said, breaking the 'Pilot, EW' communications protocols. "Your brother went to HQ from here, right?"

"Stephen," Ernest said. "He actually works in CINCSAC's office. Talks to the big man daily."

"Maybe he can tell CINCSAC how stupid all this is."

"Did you know Stephen?" Ernest asked.

"A little."

"Stephen is a lifer," Ernest said. "He's not going to tell CINCSAC something CINCSAC doesn't want to hear."

26. Renewed Hope

Omaha, Nebraska

Stephen arrived at the SAC Operations main briefing room a little early, wanting to make sure he got a seat. It was a last-minute change, but all of the primary staff was already seated. There was some light-hearted discussion which took Stephen by surprise until he saw the scorecard on an easel: no losses in the first wave and only one B-52 with any significant battle damage.

General Ham arrived and the entire staff erupted in applause. Ham ignored the reception and took his seat, scowling at the butcher paper scorecard. "Tell me about Ivory Zero One," he barked.

"Helluva good pilot," General Stringfellow said. "They had two engines out and lost the cables to their rudder and part of their elevator. They made it to their emergency field, a U.S. Marine airfield at Nam Phong, Thailand. They lost all electrics and their stabilizer trim on short final. It is almost impossible to land the -52 without the stab trim, but by God they did. He got the airplane stopped on the runway. She will fly again."

"What else?" General Ham asked.

"We finally took out Hanoi Radio," General Stringfellow said. "I know that was a high priority. All things considered, so far, it has been a good night for SAC. The third wave should be hitting the target in about an hour."

Stephen checked his watch and the three clocks on the wall, one for "Zulu" or Greenwich, England, one for Omaha, and the third for Hanoi. It was 1600 in England, 1000 in Nebraska, and 2200 in North Vietnam.

"What about night three?"

"The staff believes the plan is working," General Stringfellow said. "We just need to update the target list. The wing commanders at U-Tapao and Andersen are still complaining about the last-minute changes. The operations staff agree that has been a problem and feel that keeping to the plan will make it easier for everyone concerned. We would like to get to that right now."

"What about the full moon?"

"Yes, general," General Stringfellow said. "We have the head of our weather division here to fully brief you on that when you are ready."

"Let's do that now," General Ham said.

Major Bellum checked his watch again. It was coming up on 11 a.m. in Omaha, which meant it was nearing 11 p.m. in Saigon. He had a call to make. One of the challenges of conducting a war quite literally on the other side of the globe was the need to coordinate these kinds of calls.

Stephen punched in the necessary numbers on his desk phone to access the Defense Department's satellite phone system. After hearing the familiar dialing tones, a female voice answered. "U.S. Embassy, Saigon, to whom may I direct your call?"

"Major Haverill," Stephen said. The female voice was replaced with another tone and then a male's voice."

"Major Haverill," he said.

"Jack, it's Stephen."

"Ah good to hear your voice. How is life in the Ivory Tower?"

"Afternoon tea and scones. You know how it goes."

"Not any more I don't," Jack said. "I got something for you."

"Okay. It's got to be important if it keeps you out of the bars this hour."

"Well, it is. We passed this up through official channels and yet nothing has changed. Somebody is stopping this from getting to your boss. CINCSAC needs to know this."

"He's in a briefing right now. Maybe I can get it to him."

"We got this from a source in the North Vietnamese government," Jack said. "They figured out how to target our bombers without exposing their SAM radars until the last minute.

Now I'm not a bomber or fighter pilot so I don't know a lot about this. But I heard they normally need three radars to aim their missiles and they have to do it long enough to do the math. That exposes them to our fighters and usually ends up with one of our missiles right up their asses."

"HARM. Our Wild Weasels can shoot an air-to-surface High-speed Anti-Radiation Missile that homes in on their radar."

"If you say so. But the fact we fly the same damned route and the same damned altitudes means all they have to do is wait for the first bomber and then target the rest of the formation only turning their radars on in the last moment."

"That seems pretty obvious, now that you put it that way."

"And in whose military brain does it make sense to send the bombers in single file, one after another? I'm an airlift pilot and even I know that is suicidal."

"You say you passed this information up the chain?"

"A month ago. When we noticed the October attacks were carbon copies of the April attacks, we started digging. We pushed this through official channels. When night one of this current op repeated the pattern, we went directly to the SAC Operations Staff. I hear tonight's attacks are the same old, same old. This is making us in the Air Force look like idiots."

"Well, it can't be too bad. First wave reports are looking good. I hear we got Hanoi Radio."

"Nine minutes," Jack said. "They were off the air for nine minutes. How many tons of bombs did we drop for the sake of nine minutes? Were any of the airplanes shot down for the sake of nine minutes? You got to fix this."

"I'll see what I can do."

It was 1500 before Stephen and Colonel Snedeker had a moment alone in their office upstairs. "Second wave reports are in," Snedeker said. "Our very good day continues. No losses."

"A friend of mine in the Saigon embassy told me a report of theirs is being ignored," Stephen said. "I got a copy from the communications staff. CINCSAC needs to see this."

He handed the three-page document over and Snedeker read it. Twice.

"We can't go to CINCSAC with this," he said, handing the report back. "It is a breach of the chain of command."

"We are sending our bombers on suicide missions."

"They are hardly suicide missions. Our loss rate was three percent on night one. If the third wave today is as good as the first two, that will cut our loss rate in half. It's hard to argue with success."

"I think today might be just because the bad guys have run out of missiles."

"Ridiculous," Snedeker said. "Intel said they've got hundreds, if not thousands."

"I spent some time at the intel shop this morning. They intercepted multiple transmissions of SAM sites pleading for resupply. They say it takes hours to assemble each missile and then hours more to get them loaded. On night one they salvo'd three SA-2's per bomber, which is standard Soviet protocol. That made jamming them that much harder. So far on night two they only salvo'd two per bomber and that made it easier for our EWs to defeat them."

"What else have you got?"

"We dropped 2,900 bombs, lost one B-52, and damaged two others taking out that radio station. And those bombs could have been better spent attacking SA-2 storage facilities."

"Anything else?"

"My brother thinks we would cut our exposure in half if we didn't egress to the west. He says flying to the Gulf of Tonkin cuts our SAM exposure and gets us to possible rescue if we needed it."

"So now he's a navigator too?"

"He's pretty damned smart. He turned down an appointment to the Academy and went to MIT instead. He volunteered for nav school because he thinks pilot training is for dummies. He maxed out nav school before he got plucked out as an EW. He knows more about navigation than any navigator. After he told me about egressing east, I looked at a map. It seems pretty obvious to me now."

"So, he is a better war planner than the entire SAC staff. We should get him to plan the next war."

"I didn't mean that, sir. I think it is an option CINCSAC should consider."

"You may have noticed General Ham is a deliberate decision maker," Snedeker said. "Plus, he has no background in bombers. The other generals talk about having to manage him. It must be their considered opinion that giving him this information will not be beneficial. We have to trust that the brain trust that is the Strategic Air Command Operations Staff, is smarter than you give them credit for. Be patient, Stephen. A day from now we may find out that the plan is a good one. Our POWs could be home before the year is out."

27. Complicit

Berkeley, California

The relationship was over, Martin realized. It would be his last look at the utter chaos of a woman's dormitory room, with the combined clothing, makeup, teenager girl turned college woman paraphernalia of two women in one small room. But only one woman for now.

"She's gone home," her roommate said. "She had her last exam and wanted out of here. She left you a note."

Her roommate handed Martin the note and walked to the door. Martin followed, exited, and she closed the door behind him. The note left little to the imagination. "I cannot date someone who does not oppose this war as much as I do. Don't call me again."

Standing in the hallway of a woman's dormitory would have made him feel self-conscious not too long ago. Now all he could do is reread the note. He replayed their last meeting in his mind. She wasn't persuaded that he agreed the war was immoral, that his brothers were war criminals, and that even considering letting them conscript him into the Army would make him a war criminal. He had said all the right things. It must have been as Gene said: he gave off the wrong vibe.

As he left the building, he looked up to see Eli Saunders, one of his floormates at his dormitory. "Visiting Sherrie, I see," Eli said.

"No, she's left for home already," Martin said.

"The carefree life of a sophomore. We have graduation hanging over our heads."

"And the draft. Well, at least for me."

"Yeah, sorry about that. It is amazing how important the day you were born has become for guys our age."

The evening had turned brisk, reminding Martin of the change of seasons at home back in Omaha. The heat was over, the cold was yet to come, and the days between would make up for the ills of both seasons. It should have been a happy time.

"Do you think about the war at all, Eli?"

"I do. I am sure I don't as much as you. It impacts you more than me, obviously. The news has been bad, hasn't it?"

"Yeah. Strange how there haven't been any protests. It seems things over there are as bad as they ever have been. But over here? Nothing."

28. The Other Side of Confusion

Hanoi, North Vietnam

"Our allies claim many more bombers hit," Sergei said. "But I find that unlikely as we have no reports of new wreckage anywhere. It is my estimate that we did not get a single bomber last night." The captain seemed to be weighing his words carefully, staring at his own feet while standing at the correct distance when addressing a superior officer. But then he seemed make a "to hell with it" decision. "It was a failure," he finally said.

"Was it the American tactics?" Nikolai asked.

"No, the tactics were identical," Sergei said. "I am afraid it was the chaos created by our allies and their reorganization of their command and control."

"In the middle of a bombing campaign?" Nikolai asked. "Whose idea was that?"

"I do not know. They said it was necessary to improve on the triangulation process. They thought reducing a layer of the communications would speed the calculation process and I think that is correct. The reorganization is going to pay dividends. But the result last night was missile commanders without precise targeting instructions."

Nikolai sat at his desk, attempting to lean back on his rickety wooden chair, but the lumber's protest made him rethink the idea. The night had been terrifying again, but only because of the incessant noise and shaking of the earth. He chose quarters in one of the Prisoner of War camps, knowing he would be safe. But even expecting all manner of hell three times every four hours wasn't enough to allow him to sleep.

"You said the tactics were identical?"

"Yes, major. You could almost set your watch to the strike of the first bomber. I am starting to think we have been overestimating these Americans."

"What of the jamming?"

"Again, it was identical."

"I have been thinking," Nikolai said, drawing a lopsided triangle on the dust of his desk. "The cloud of noise disguises three bombers in a triangular formation which appears to elongate as they approach their target. Why is that?"

Sergei picked another spot of the desk to draw three dots in a triangle. He then picked another spot and drew three dots in a row. "The geese getting into single file to drop their eggs?"

"Perhaps, hard to say. The noise is just too large. But notice that the noise, ah, the jamming decreases at the front of the triangle as they drop their bombs."

"Could it be they cannot jam and bomb at the same time?"

"I cannot believe they would design such a bomber. But it appears possible. But that is not the point, Sergei. The point is we know inside this cloud of noise there will be three bombers."

"But the size of the noise cloud is too large to target and hope for a hit, major."

"It is at first. Let me use the photos instead." Nikolai retrieved the three photos of the April bombing and arranged them in order of time. "You are right that as the bombers first approach, the cloud of noise is too large. But notice that after a minute, the cloud is much smaller."

"I see that. Perhaps we can target the middle of the cloud, and uplink a more precise target location after missile launch. That gives us a much higher probability of a good target solution."

"But there is something more important that has eluded me," Nikolai said. Sergei looked confused for just a moment and then his face brightened.

"Yes, of course. We cannot burn through the larger cloud. But the smaller cloud?"

"Ah, I had not thought of that. Good. One more tactic to consider. If the cloud of noise elongates, we can better guess the position of the last bomber."

"Could the Americans be so foolish as to repeat the pattern for a third night?"

"It is unlikely. But we cannot discount such foolishness. Let us spread these ideas to our allies."

29. Linebacker II, Night 3

U-Tapao, Thailand

"Well, at least we're consistent," Major Stevens said as they boarded their crew bus for the third day in row. "I'll make this short and sweet. We are in the fifth cell, Orange Cell, and we are number three."

"I hate being number three," Ernest said.

"Me too," Kevin said.

"And let the bickering begin," Stevens said. "Just study your packages and let me know if you see anything different."

"Hey, we get to hit something other than a radio station," Kevin said. "It's that railroad yard on the northeast side of Hanoi. I saw it on the list yesterday."

"And the day before," Tommy said. "You notice SAC is sandwiching the Andersen birds between us U-T birds today?"

"Yeah," Harry said. "So far Andersen has lost three Buffs to our zero. Maybe they think our luck will rub off."

"Or maybe our skill," Stevens said as the crew bus pulled up to their aircraft. "Well, let's get the show on the road."

Ernest kept his thoughts to himself until Orange one and two were airborne. "On time," Ernest said. "That's something new."

After level off, Ernest unstrapped and went forward. He knelt on the cockpit floor to look out. "Another full moon," he said. He wasn't plugged into the interphone and the pilots didn't hear him. The moon shone on the cloud deck below. He had heard about the line of rotating beacons on the previous night, but now he saw them. But even without them, he could see the silhouettes of lead and number two, right where they should be.

"Beautiful night," Harry said, looking aft. "Makes you appreciate things, don't it?"

"It does," Ernest said. "How is it, Harry, that you never get scared?"

"Never get scared? What are you talking about?"

"I've never seen a SAM with my own eyes," Ernest explained. "I've lost count of how many we've had headed right at us. And you just calmly call them out."

"Oh," Harry said. He looked at Major Stevens in the left seat and then back at Ernest. "I am plenty scared. But I got to have faith in the jet, in my crew, and in my God. I have faith in you, Ernest. We are going to be a-okay."

Ernest nodded his head and returned to his seat. As he plugged his helmet back into the oxygen and communications line, Kevin was speaking.

"I know you got to do what you got to do to maintain cell position, pilot. But we need to be centered down here or we can't drop."

"Yeah, I know," Stevens said. "But when you start on the left and lead turns right, something has to give."

"I'm just saying we need to be on course by the time we get to the target. That's all I'm saying."

"I hate being number three," Ernest said to himself again.

"It's quiet on the radio tonight," Harry said over interphone.

"The guys are getting used to being shot at," Major Stevens said.

We got a light show off our right, pilot," Sergeant Polanshek said.

"Roger that," Harry said. "I see some secondaries, the guys got something important, I bet."

"Raise your eyes," Polanshek said. "About our altitude over the secondaries."

"Oh," Harry said.

"What?" Kevin and Tommy both asked from their windowless basement.

"Somebody's been hit," he said. "Big explosion and parts falling to the ground."

"IP in ten minutes," Kevin said.

Ernest rechecked his largest screen, the ALR-20 receiver. The activity at a distance was on a par with night two, but much less than night one. But there wasn't as much noise coming from the cell. Something wasn't quite right.

"Pilot, EW. Are we in position?" Ernest asked.

"Sure are," Major Stevens said.

"Jamming from the cell isn't as heavy as it should be," Ernest said.

"Oh," Stevens said. "I think I heard Orange Zero Two say they were down three jammers."

"That means we are down three jammers," Ernest said. He felt the aircraft bank to the right and watched on his scope as the formation appeared to move further to the right.

"IP," Kevin said. "I need three more degrees right, check your PDI, pilot."

"Sorry," Major Stevens said. "Big crosswind tonight."

Ernest thought about the crosswind, the turn to the right for the bomb run, and the number three. But it was a thought that would have to wait.

"Sixty seconds," Kevin said. "Still not on course."

The collective jamming noise from the cell was considerably lower. Ernest, for the first time, realized the decrease was caused not only by Orange 02's reduced number of jammers, but the fact his aircraft was so far out of position. Then came the rattler.

"Fan Song," Ernest said. "They are looking at us."

"Thirty seconds," Kevin said. "Doors. We are just about on course."

"Uplink," Ernest said. "One ring. Fangs."

"Lead's in the turn," Harry said. "Time to drop?"

"Not yet," Kevin said. "Twenty seconds. Stop maneuver, level."

"Launch!" Ernest said. "Three and a half rings, headed right for us."

"Pilot, gunner," Polanshek yelled, "two SAMs right under us!"

"Not yet," Kevin said, "five more seconds."

"Bombs . . . " Kevin said but his voice and all other sound was eclipsed by an explosion aft of the crew compartment. The sound wasn't what Ernest was expecting. It was louder, closer, and somehow, more metallic.

Ernest felt himself tumble forward and pulled his head away from the scope hood just in time to see it vaporize. He reached down for his ejection handles and everything changed in an instant. The world went from dark and hot to becoming both very bright, and very cold.

30. Delayed News

Omaha, Nebraska

Stephen timed his drive so as to arrive at the building at 0700, which gave him time to grab a coffee, make it to his office, check his inbox, and then get ready for Colonel Snedeker's staff meeting at 0830. The first TOT would be at 0800, Omaha time, but they wouldn't be getting any solid intel until around 0930. The first U-Tapao bombers would be landing around 1100, but they were instructed to pass coded messages as soon as they made it out of the combat zone.

"Night three!" Snedeker said to his staff. "Let's hope it is as good as night two. CINCSAC's meeting at 1000. Stephen, I need you to skip that. The plans people are doing a cut drill on operations funding, something about what we are expecting with the new Congress next month. I need you to sit in on that, make sure they don't get crazy. General Ham tends to give on the budget when the pressure comes from the Defense Department. We might as well be ready for that. Report back to me at noon."

Stephen felt the hours watching over the budget were well spent and was eager to report back to Snedeker. He entered the office and immediately detected something weighing heavily on his boss, who rose, gestured that Stephen should take a seat, and then closed the door.

"Your brother's airplane was hit," he said. "I'm very sorry."

"Chutes?" Stephen asked immediately, even as he felt the blood leave his head and his spine lose all rigidity. He slumped in the chair.

"None seen," Snedeker said. "The explosion was seen eighty miles away by a reconnaissance airplane. Death must have been instant."

Stephen's first thought was a vision of Ernest's Air Force commissioning. His father pinning one of Ernest's gold bars on one shoulder while he did the honors for the opposite shoulder. Dead?

"You should take some time, Stephen," Colonel Snedeker said. "Your parents will be notified through official channels, but they should hear it from you. They still live in the local area, right?"

"Bellevue. But I need to know more about how . . . more about how he died."

"Are you sure?"

"Yes."

"We know more than usual," Snedeker said. "Your brother was in number three, wings level, bomb bay doors open. Number two was in their post target turn and their gunner spotted the SAMs and called them out. He was able to follow the missiles all the way into your brother's airplane's bomb bay. The explosion was so bright it blinded him for a moment. The next thing he saw was fragments of the airplane falling to earth. I am so very sorry."

* * *

Getting out of his car, standing in his driveway, Stephen realized he had no memory of leaving Building 500, getting into his car, or the drive home. He stood for a moment, looking at his front yard that had just gotten a light dusting of snow. He needed to sit, so he sat. The snow was soaking into his pants, he thought.

The front door to their house opened. "Stephen?" Susan ran to his side and knelt to look him in the eyes. "Stephen, what's wrong? Talk to me."

"Ernest is dead," he said. "Shot down. He's dead."

She fell to her knees and wrapped her arms around him. She started to cry softly and then came the uncontrolled sobs. Her tears are staining my shirt, he thought. But that's okay. She will cry for us both. When she was cried out, he looked up and realized his vision was blurred by his own tears. "We need to see my folks," he said.

"Yes."

"Here, take my hand," she said. "Let's get you dressed." He followed her into the house and into their bedroom. "He was shot down?"

"Yes."

"Do you think he . . ."

"Suffered?"

"I hope not."

"No, they said it was a missile to the bomb bay. It must have been instantaneous."

She drove them the twenty-minute distance to Bellevue. He glanced down at his black pants and gray shirt. Did he change clothes? He should drive. No, she's driving.

* * *

They are sitting. They are looking.

"Stephen, talk," his mom said. "You are scaring us."

"It's Ernest," he said at last.

"No!" she said. Susan came to her side.

"He's . . ."

"Gone?" his dad said.

"Dead," Stephen said. "Ernest is dead."

"No!" his mother screamed. She fell into Susan's arms, crying. Susan's sobs started again, and Stephen started crying too.

"Gone," his father said, quietly, his eyes closed. "A missile?"

"Yes," Stephen said. "Over Hanoi."

He sat alone in his chair, opposite his father. Stephen had no words left. But his father had no words at all.

31. Alone

U-Tapao, Thailand

After three nights the routine was feeling almost normal. Marjorie didn't expect the trauma of the morning after Night 1. Ernest never called after returning from the first night's mission and she didn't hear about the number of downed airplanes until she showed up at the weather shop at noon.

She had an anxious Night 2 but managed to fall asleep. The next morning, Ernest's note settled her nerves magically. She was hoping for the same result after Night 3.

She woke up early, still anxious, and went straight to the door. There was no note. Ernest had left his Night 2 note folded and placed in the metal label holder which held her name on a card above her room number. She removed the card, perhaps the note fell under it. No.

She stepped outside in her bare feet, still dressed in her pajamas, looking left and then right. The palm trees surrounding the building were swaying gently in the wind. Perhaps the note had blown off the door. Nothing. It was 0500, too early to call. He will have been exhausted after his midnight flight. She owed him his rest.

She put on her fatigue pants without the belt and the olive T-shirt that went underneath her uniform top; they had grown into the most comfortable clothing she had ever had. She checked her closet and confirmed she had at least one uniform shirt ironed and ready to go and considered getting ahead of her ironing with the few that awaited that chore. She needed something to do. Something else.

She paced the room. The evening flights meant she didn't have to report to the weather shop until noon but going back to bed was unthinkable. She stared at the hands of her room clock, mentally urging them to speed up. At 0600 she would call. She poured herself a bowl of cereal and topped that with milk. It tasted like wet sawdust in her mouth. She couldn't bring herself to eat another spoonful. Once the minute hand was straight up, opposite the hour hand, she dialed. No answer.

She ran to his room in her bare feet and pounded on the door. "Ernest!" Over and over. "Ernest!"

Several of Ernest's neighbors opened their doors, looked, and closed them quickly. Marjorie looked at the few who didn't retreat. "Have you seen Ernest? Captain Bellum?" The looks on their faces told her all she needed to know. Her knees buckled and she fell back onto Ernest's door, sliding along its smooth surface until she was crumpled on ground. She held her head with her hands, propped on her knees.

She was aware of her own cries of anguish, the sobbing, and spectacle she must have been, sitting on the ground near Ernest's door. But there was nothing to be done about it. Her shirt was soaked but had turned cold. It was very cold. And there was speaking.

"Ma'am?"

She felt totally aimless. What should she do? What could she do?

"Ma'am, can you hear me?"

Of course, she would have to get up. She had to press on. But why? She felt the cold of the concrete beneath her legs. She felt cold.

"Address her by rank and name," another voice said. "That will usually startle them out of it."

"Lieutenant Ralston."

She looked up. There were two orderlies dressed in the blue smocks reserved for nurses, doctors, and technicians at the base hospital.

"Lieutenant Ralston," one of them said, "you need to come with us."

"Okay." She rose unsteadily, prompting each orderly to grab an arm and help her to her feet.

"Where are your shoes?"

"What?"

"Your shoes? Where are your shoes?" She looked down at her feet, still bare and a bit blue from the cold.

"In my room, it's the next building. Can I get them?"

"Of course, ma'am."

32. Nicholas

Hanoi, North Vietnam

Sitting on the dirt floor, alone in his cell, Ernest tried to assess his situation though he had difficulty in tracing the steps that led him into the cell in the first place. His arms were tied behind his back at his elbows so hard, he thought they might pop out of their sockets. His vision was blurred in one eye, but he could see clearly with the other. Light shone through a small window mounted high in the wall. He took a quick inventory. He was intact, no missing parts that he could see. The bottom of his flight suit was charred black, and parts of his boots were singed. The smell of burnt rubber suddenly hit him. His charred flight suit legs alerted him to another pain; his legs felt sunburnt.

He was alone in the cell. "Make use of the time," he thought. But how? In basic Prisoner of War training, he was taught how to resist interrogation but also taught that he had a breaking point. Expect it. But don't make it easy. Resist. In advanced POW training, the one reserved for EWs, he was taught there are things worse than torture and that the only survival technique was to avoid being found out as an EW in the first place.

The walls have ears, he was told in advanced training. EWs were frequently screened and any tendency to talk in one's sleep was a quick way to lose a security clearance. Ernest never had that problem, but he was prone to talking to himself, a way of sorting out problems. *Don't do that,* he reminded himself.

His immediate problem was being discovered as an EW, that would mean a one-way ticket to Russia. *I am a navigator,* he thought. *My job is Point A to Point B. That is what I do. I sit in the right seat, working with plotters and a compass.*

But what if they had Kevin and Tommy in custody? What then? Or worse yet, what if one of the pilots or the gunner gave the names of the real navigators? What then? In advanced POW school he remembered a moment of panic when his fake interrogators started roughing him up, pumping him for the identity of his pilots. Once they got that from him, they claimed to have beaten his identity out of the pilots and all was lost. His school captors were not allowed to inflict lasting damage on their students. The pain in Ernest's shoulders reminded him the stakes now were much higher. *Think!*

Ernest thought back to his days as an EW trainee assigned to a student B-52 crew. While just a few years ago, he remembered everyone clearly. He had been assigned to several crews since, but he would always remember his first crew.

I am Captain Ernest Bellum, navigator, he rehearsed in his mind. *My aircraft commander is Major Allen Boyd, a little dictator with a short fuse who shoots first and aims second. My copilot is Lieutenant Calvin Perry, a nice guy but inexperienced. My radar navigator is Captain Scott Nelson, a really smart officer who watches over the navigator. Who watches over me. The EW is Lieutenant Teddy Wilson, a very smart guy who bypassed navigator school because he's too smart for that kind of thing. No, let's make Teddy a captain. The gunner is Staff Sergeant Gary Henderson. I hardly know him, but he seems okay. I don't know anyone by the names you mention. Stevens? Broadmore? Never heard of them.*

There was more to it, he knew. You can't expect them to take your word for it. You needed a persona. A quirk. A distraction. Ernest realized that he had the perfect persona ready to go.

There was a click at the door, and the shuffle of feet. The door opened and a guard entered. He was short but muscular and wielded his rifle with the stock pinned between his hip and his right arm. He motioned with his left; Ernest struggled to get to his feet. The guard lifted him by his hands, still tied behind his back.

"Ahh!" he yelled. The guard pushed him into the wall. Ernest felt cold steel between his arms but then realized a blade was slicing through the ropes. His arms sprang forward, causing more pain and more tears of agony.

The guard pointed to the door with the rifle. Ernest found it difficult to walk at first, his left knee buckling after the first step. He found his footing as they left the cell, and each step became easier. The guard said something he didn't understand and pointed to the right. Ernest turned right and kept walking.

Like his cell, the path alongside the building was compacted dirt. There were other doors along the path, but nothing to distinguish one cell from the others. He tripped on his own feet and fell into the building, causing the walls to shudder. The guard said something but waited for Ernest to pick himself up. It was a wooden structure, single story, with a small courtyard with bushes and what looked like fruit-bearing trees. The vision from his bad eye was returning and with it some depth perception. The courtyard was lined with a wall, a high wall. There may have been barbed wire, but when he turned to look the butt of the guard's rifle met his head abruptly, knocking him to the ground. He looked up to see the guard smiling, enjoying the spectacle. Ernest got back on his feet, thinking, *I dub thee Satan.*

They came to a room at the end of the building and another guard gestured him in and pointed to a chair positioned next to a small table. Ernest sat. The new soldier said something, and Ernest repeated his name, rank, and so on, preparing himself for another slap to the head. But it didn't come. The soldier just looked at him blankly. This guard was half the size of the first, wore what looked to be a green uniform with no visible insignia except a single red rectangle with a gold stripe in the middle. A private, Ernest thought. Barely old enough to shave. *You, I shall call Pee Wee,* he thought.

Ernest massaged his elbows and looked past the guard. He focused on infinity and willed his expression to become blank. It was hot and humid, much worse than U-Tapao. *This place could use some air conditioning,* he thought, causing a smile that he immediately stifled. *You have a role to play.*

"Hey, E. W." a voice said from the door. Ernest continued to look ahead, only turning once the individual entered the room. He was a tall Caucasian man with blond hair, dressed in an orange shirt with yellow stripes, slacks, and canvas shoes. "I am Nicholas. I am wanting to see how they are treating you."

Ernest continued to look ahead. Nicholas pulled a chair opposite Ernest and smiled. "I am an outsider; I am sure you can see. How long have you been here?"

"Nicholas?" Ernest asked, trying to appear bewildered while remaining somewhat disinterested. The man's voice wasn't American, but he was speaking English. There was an accent Ernest couldn't place but he had a suspicion. He pronounced the first syllable of his name without the American sound "ih" but more of an "ee" to it.

"Yes," he said. "I am an electrical engineer doing some study here," he said. "I am big fan of all things electrical engineering. How about you?"

Ernest remained silent, willing his face muscles to remain relaxed and his eyes steadily forward. "I have never met an American EW," Nicholas said. "Such a big airplane, it must be quite a challenge!"

"Yesterday," Ernest said.

"What?" Nicholas asked.

"I got here yesterday," Ernest said.

"Oh, yes, of course," Nicholas said. "I was saying that I have a hobby. I study all things engineering. Such as your B-52. An impressive machine. Yes?"

Ernest shifted his eyes towards Nicholas for the first time. Nicholas smiled, exposing the eyes of a chain smoker and a gold tooth. He was tanned, but not with the darkened skin of an Asian, more of a beach going touristy type of tan. "Yes, very impressive," Nicholas said.

"Treating me," Ernest said.

"What is that?"

"Treating me," Ernest repeated.

"Yes, of course! Perhaps some food and water? Yes?" He gestured to the guards. One of the guards left and returned with two bowls and two cups. Ernest watched as Nicholas used the spoon in his bowl to devour the food in three or four mouthfuls and then eagerly drank from his cup with loud gulps. He finished with a guttural "ah!" Ernest cautiously picked up the cup took a tentative sip and then gulped the rest of the water. The other bowl had rice and vegetables he didn't recognize. He took a mouthful and chewed gingerly and then ravenously.

"Now we can talk," Nicholas said.

"Captain Ernest Bellum," Ernest began but stopped when Nicholas raised his right hand in a stopping motion.

"No need for that," Nicholas said. "I will do the talking, yes? I am an engineer, always looking to learn. We engineers are like that, yes?"

"Rice," Ernest said.

"What?"

"Rice," Ernest said.

"Yes, yes," Nicholas said. "More later. But now I want to talk about engineering."

Ernest returned his gaze to the spot on the wall he had noticed before. He willed his eyes to glaze over.

"Me, I prefer the radio," Nicholas said. "I study all things to do with radio. A.M., F.M., H.F. I like them all."

"Thank you," Ernest said.

"What?"

"The rice."

"Ah yes, of course. As I was saying, which form of radio waves do you prefer?"

"Navigator."

"Ah yes," Nicholas said. "We engineers are different. The soldiers, they do the killing and bombing and all the, ah, distasteful business. But we deal with electricity, radio beams, and things they do not understand. The war will be over soon, and you will be going home. So, I have limited time to understand something I have been puzzled by. When your B-52s come to drop the bombs, there is much noise in the air, the jamming of the radio waves. But as soon as you drop the bombs, the noise stops, and these savages have an easy time tracking you."

"Navigator," Ernest said.

"Yes, so you say," Nicholas said.

"April 10th, 1947," Ernest said.

"What?" Nicholas said.

"Captain."

"As I said before, there is no need for that."

"Keep us on course."

Nicholas, still smiling, looked at his watch. "I will be leaving now. I don't want to be here tonight when another B-52 falls on the city. I hope you survive so we can talk some more tomorrow. I will mention to our friends here your fondness for the rice."

Nicholas spoke and the door opened. Nicholas left and Pee Wee returned. The guard gestured to Ernest with his rifle, pointing to the door. Ernest retraced his steps but didn't remember which cell. When he passed it, the guard spoke again and opened the door. Ernest entered and saw that he now had a bucket and another cup of water. The smell of the bucket was unmistakable, as was its purpose. First, some housecleaning. He placed the bucket at one corner of the cell and his cup of water at the opposite corner.

He sat and replayed the meeting in his head. Was his act convincing? "Nee koh lass" was obviously Russian. The Soviets were eager to learn all they could about the B-52's ECM suite. They could not have known that he was an EW. His flight suit had no patches that would have given that away and nothing on his person should have revealed that. Was his act too obvious? Was saying anything at all a violation of the code of conduct?

Ernest had a good memory, exceptional, in fact. "I am an American, fighting in the forces which guard my country and our way of life. I am prepared to give my life in their defense." That much he remembered for sure. There was something in there about not divulging any information. "When questioned, should I become a prisoner of war, I am required to give name, rank, service number, and date of birth. I will evade answering further questions to the utmost of my ability." Ernest concluded that his abilities were meager at best. His arms hurt and he was still hungry. Hardly torture by any measure.

"Critique," Ernest said aloud. The only way to improve at something is through criticism, he had always told himself. His job now was to be a POW and the stakes were higher for him than any generic soldier. The instructors at POW basic would be aghast at his performance with Nicholas. They would have critiqued him for offering "navigator" and the fact he had been shot down just a day prior. His performance would have earned him an "F" if such grades were given. But his advanced POW instructors would have given him an "A," he thought. Plant the seed, but don't overplay it. Make Nicholas feel that he was clever and had extracted the fact that Captain Ernest Bellum was a navigator, an officer of little interest.

"Anticipate . . ." Ernest said to himself but then stopped. *Stop talking to yourself,* he thought. *Anticipate the bad guy's actions, act on that, don't react without thought.* That was one of his father's many pithy sayings that seemed meaningless at the time, but suddenly useful. If Nicholas knew Ernest was an EW, he would have been packed up for Russia already. But they don't want to waste the effort on a mere navigator, *like me!* What would Nicholas' next move be?

Nicholas will try to trick me. If he has the real crew, he'll use their names to convince me he knows who I am. If he doesn't have them, he'll say he does without using their names.

Ernest tried to lay on his back, but the cold floor on his injured shoulders was just too painful. He leaned against the wall gingerly, rehearsing the names of his old crew and thinking about their personalities. Major Boyd was a little Hitler, always finding faults in others. There was the time he exploded at the navigator for not ensuring the crew entry hatch was sealed. It turned out to be a problem with the handle, but Boyd just couldn't let it go. The navigator was quaking in his boots that day. *I was quaking in my boots that day,* Ernest thought.

"Major Boyd, I closed the hatch!" he pleaded aloud. Then, with a whimper, "I promise you, the hatch is closed." *Don't oversell it. You can mumble it tomorrow, but that should do for now.*

He drifted to sleep thinking about *her* and how their height difference allowed her to place her head right on his shoulder, so he could take in her smell. It was almost as if by design. He asked her what the delightful smell of her auburn hair was, and she laughed that laugh of hers. "Head and Shoulders dandruff shampoo." His hands were just the right size to encircle her arms when he pulled her near. She was almost spindly, breakable, and delicate. She needed his protection. He thought of the anguish in her eyes following the first night of the mission and how he had been careful to write her a note after the second night. He wrote, "I'm okay, see you tomorrow," and left the note on her door. He was tempted to write, "I love you," but didn't. And by now she had discovered there wasn't a second note waiting for her. And then the earth shook.

The bombs were falling on Hanoi again. He opened his eyes. The flashes came first, then the explosions. It was much like a lightning storm and thunder, only more so. The noise was almost as intense as the last sound he heard in flight.

He relived those moments, searching for clues. Kevin complained about the crosswinds and said "not yet" to Harry's pleas to drop their bombs early. He thought of his arguments with Major Stevens about having to open the bomb bay doors sixty seconds early and his complaints to his older brother. SAC changed the ROE for the doors to thirty seconds. Was that Stephen's doing? He thought of his ECM panel vaporizing in front of him. If the blast was in front of him, the pilots and navigators would be doomed. *No, don't think that.* If he made it out, surely the others could have too. But he was watching the missile on his scope just before it hit; he had the best view even though he didn't have a window. Did he eject too soon? What happened to his crew?

33. Grieving

Omaha, Nebraska

"What are you doing here?" Colonel Snedeker asked. "Shouldn't you be at home with your parents?"

"There is only so much grieving to be done," Stephen said. "Jonathan, dad, he just kind of sits and keeps to himself. He was in a car accident recently; I think maybe it takes a while for all this to sink in. My mom's sister came and the two of them cry all day. I gave my brother in California a call, but he doesn't want to talk. Susan, my wife, just barely knew Ernest. But he's the first person she's really known that has been killed. There wasn't a single death in the squadron at U-T while I was there. But there is nothing for me to do at home but feel bad. I thought I could make use of the time and keep my mind busy."

"Fair enough," Snedeker said. "The last day has been a nightmare. Here's my log," he said, handing over a bound ledger. "This is between you and me, don't breathe a word of it to another soul. This diary doesn't leave this office. It will take you a while, I'm needed downstairs. I'll be back in an hour."

Stephen sat back in the chair, as Snedeker left his office, closing the door behind him. The entry marked December 20, 1972 was twenty pages long. The first page was labeled "First Wave." The first paragraph was titled Quilt 03 and detailed how the Andersen G-model was hit during its post target turn, with the phrase "possible survivors" underlined. The next paragraph was dedicated to Brass 02, another Andersen G-model. The phrase, "bailed out, crew recovered" was underlined. Then came Orange 03 with "no survivors" underlined with the words "Bellum's brother?" circled. Stephen stared at the text that followed, sobbing gently at the description of the fireballs seen by aircraft in the area. He got up and looked outside Snedeker's office window at the frigid Nebraska landscape.

"Move on," he said to himself, picking up the notebook where he left off.

The page titled "Second Wave" was empty except for the words, "called back" and another entry written in a different ink. "Operations Staff irate."

The page titled "Third Wave" was a return to the form of the first. Straw 02, an Andersen D-model, hit after bomb release, "bailed out, one MIA, five recovered." Olive 01, an Andersen G-model, hit during the post target turn, "no chutes, survivors doubtful." Finally, there was Tan 03, another Andersen G-model. "Not sure why" was underlined, followed by "crew fate unknown."

Stephen quickly tallied the results: 6 B-52s hit by SAMs, four of those were G-models, all but one from Andersen Air Base in Guam. He turned the page, expecting a blank, but there was much, much more.

"Peace talks broken off," followed by "they think they've won."

The next page was titled: "17th Air Division Commander Off Reservation." Brigadier General Silvers sent a message to CINCSAC, outlining a new tactic. He didn't ask permission of his immediate boss, the 8th Air Force Commander. Instead, he summarily informed HQ SAC that this was how his crews were going to fly future missions.

"Whoa!" Stephen said aloud. This was a good way to end a career, he thought.

"B-52 crews will no longer fly the same routes and altitudes with staggered times."

"Will attack from multiple directions separated by varying altitudes."

"No more post target turns. Head for the ocean instead, fly south to South Vietnam before heading home to U-Tapao or Guam." It was just as Ernest had suggested.

"U-T crew used chaff. It worked. Will allow crews to use chaff at their discretion."

The next quote was from the 8th Air Force Commander: "I agree."

The final quote was from CINCSAC. "We cannot continue to jeopardize our nuclear fleet! Limit all future efforts to no more than 30 B-52s per night."

Just as he finished the last page, Snedeker returned. "What do you think?" he asked.

"A day too late for Ernest," Stephen said. "But General Silvers just saved a bunch of lives."

"I think you are right about that. The operations staff here is still at a total loss. They are only now getting feedback about the first night. That's all CINCSAC had to hear."

"What did he say to Silvers?"

"I don't know, but I know what he said to the SAC Operations Staff. He told them to hand night five over to the 8th Air Force."

34. Ernest Bellum, Navigator

Hanoi, North Vietnam

Ernest didn't have any visitors during the day, other than Pee Wee opening the cell and pushing in a bowl of rice, a cup of water, and a new bucket. He gestured for Ernest to return those from the previous day. Pee Wee's rifle was slung around his shoulder to free both hands for his menial tasks. Ernest thought briefly that if he were Steve McQueen in "The Great Escape," perhaps he could "bust his way out of the joint." But he could barely lift his arms without wincing in pain. The heroics would have to wait.

He braced himself for another night of bombing, which he figured would be night five of Linebacker II. But it never came. The only sound was from a distance, perhaps a thunderstorm.

"Pilot, nav," he said quietly while pretending to sleep. Ernest counted his heart beats, biding his time. In a few minutes, if they were listening, he could feed the fish more line. "The hatch is secure," he said. "Major Boyd, the hatch is secure."

That's enough for now, he thought.

The next morning Pee Wee appeared at his cell door and said simply, in English, "come." Ernest followed him down the now familiar path to the familiar room and was greeted by Nicholas.

"I am happy to see you, my friend," he said. "I hope you slept well. It appears your B-52s needed a rest last night. I must tell you that was good fortune for the city! I hear that they had shot so many missiles the last few nights that they ran out! It was good to have the night off so they could deliver and assemble the missiles from the warehouses!" Each sentence ended with the same high pitch, sing song note. "Now we are fully stocked again, and we should be safe for when your airplanes return."

Ernest sat quietly, aiming his unfocussed eyes on the wall just right of where Nicholas was sitting. "There is much for us to discuss today," Nicholas said. "The American Air Force has been so kind to drop a number of your B-52 bombers onto Hanoi. I am thinking of starting a collection. So, my time with you shall be limited, I am sorry to say. I have so many EWs to speak with today. You will be happy to hear we have the rest of your crew in the next compound. Everyone is well, I am happy to say."

Ernest turned his head toward but short of Nicholas, keeping his expression blank. "Rice," Ernest said, weakly.

"Major Boyd is especially concerned about you," Nicholas said.

"I slept."

"What?"

"I hope you slept well. You said 'I hope you slept well.'"

"Ah, so I did."

There was some activity outside and two guards carried in what appeared to be part of an aircraft instrument panel. Ernest tried to feign indifference, but he could clearly make out a section of a B-52 EW's instrument panel. The panel had some burn residue and the nomenclature on many of the instruments were unreadable. But the pattern of knobs and buttons was unmistakable. It was a bank of AN/ALT-6B jamming transmitter control heads, the kind used in the unmodified G-models. The printing on the control heads was unreadable from years of use, but the function of each switch and knob was obvious to a trained EW. *Of course, these panels means nothing to me, a navigator.*

"This might have been from your airplane, yes?" Nicholas said. The answer was no, it couldn't be, since the D-model had the newer AN/ALT-22. Ernest kept his expression blank.

"No?" Nicholas said.

Ernest sighed, feigning boredom.

"I am wondering," Nicholas said. "What are the functions of these three knobs?"

"Quiet," Ernest said, moving his eyes slowly from a point on the wall to his interrogator's.

"What?" Nicholas asked.

"It was quiet last night," Ernest said.

"Yes, as I said. I have a theory about these knobs. In radios we have frequency and amplitude. Yes? But what of the third knob?"

"Rest," Ernest said.

"Yes, yes, yes," Nicholas said. He was betraying some anger, Ernest thought. The pain in his shoulders and arms was still sharp, but with time it had stopped clouding his thinking. *Keep it up, but don't overplay your hand,* he thought. *They have my old training crew in custody! Those guys are still in California!*

Nicholas waited a few minutes and, seeing nothing more from Ernest, called out to the guards. The guards returned with another panel, the AN/APN-218, from the navigator's station. The guards spoke with Nicholas as soldiers would address an officer and Nicholas spoke with them as officer to enlisted. But his demeanor changed whenever he spoke to his prisoner.

"Here is another valuable find," Nicholas said. Ernest recognized the panel from his first few months in navigator school, before he was plucked out to become an EW. It was something they used in the Convair T-29 navigator trainer, as well as the B-52 or just about any aircraft with a navigator.

"Navigator," Ernest said.

"Yes, yes," Nichols said.

"We use Doppler to measure sound waves."

"Doppler!" Nicholas said. "I know Doppler. Austrian physicist, no?"

"Scott."

"What is Scott?"

"Speed and drift," Ernest said. "Very important for navigating."

"I see," Nicholas said.

"Scott," Ernest repeated. "Taught me how to use drift."

"I see, Scott is a who, not a what."

"Scott says zero the drift to stay on course, but . . ." Ernest said with a flourish.

"But?"

"But you add drift to heading when correcting to course. Very important."

"Ah," Nicholas said. "I think we have what we need." He gestured to the guard who retrieved the two panels. Nicholas stood and said something that sounded Vietnamese. Satan walked in and looked at Nicholas, who nodded. Satan grinned.

Satan approached and grabbed both of Ernest's arms, lifted him from the chair and threw him to the wall, as if throwing out dirty dish water. Ernest hit the wall right shoulder first, reigniting the agony of his first day in captivity. He crumbled to the floor, crying in pain.

Nicholas walked over and knelt onto both knees, his face just inches from Ernest's. "You are the most annoying American I have ever met," he said, dropping the sing song accent. "I grow tired of humoring you. Now tell me, who is Major Boyd?"

"Pilot," Ernest said, still crying. "He's my pilot."

"He's not your EW?"

Ernest said nothing until Satan returned and lifted him again, this time with his arms behind his back.

"Ahh!"

"I shall not ask again."

"Pilot. Major Boyd is the pilot. Teddy is the EW."

Satan released Ernest who fell to the floor again, crying. "Captain Teddy Wilson."

Nicholas looked at his broken prisoner. "How is it somebody so stupid can be a navigator? It is a wonder you could find your airplane, much less Hanoi." He shook his head and walked out.

35. A Single Wave

Omaha, Nebraska

"General, I would like to lead with the best news we've had in a while," General Stringfellow said. "No B-52s were hit, despite lots of SAMs fired and a couple of near misses."

Unlike previous meetings, General Ham allowed himself a smile and a fist pump. "Details."

The conference table was surrounded by the highest-level generals of the SAC Operations Staff, as usual, but the chairs were pushed more tightly together to allow space for a huge easel, measuring at least ten feet from end to end. A long sheet of butcher paper was unclipped and allowed to fall to the floor, revealing a map of the coastal area just east of Hanoi and the coastal city of Haiphong.

"As you know," Stringfellow continued, "it was an all U-Tapao show last night. They attacked the Port of Haiphong in a single wave, coming in over the Gulf of Tonkin and exiting over the Gulf of Tonkin. They were able to time a full force of F-4s laying down a blanket of chaff at precisely the moment of our attack."

There were smiles all around as two of Stringfellow's aides clipped another chart on the right side of the easel, revealing sorties, bombs dropped, SAMs fired and missed, and the best number of all: zero bombers hit.

"But I take it not everything is wine and roses," General Ham said. "So, let's have that too."

"Yes, general," Stringfellow said. "We got our first reconnaissance photos in from the first three days. Many of the targets we thought we hit hard, we barely hit at all."

36. The Hilton

Hanoi, North Vietnam

Ernest awoke to a sharp pain in his right shoulder and a ringing in his ears, surprised he had been able to sleep at all. He had been dragged back to his cell in agony, but conscious of a need to hide a satisfied grin. *I beat you, Nicholas. This annoying American beat you.* He felt confident for the first time since his arrival. He was going to be a prisoner in a POW camp in North Vietnam, not a science project in a lab somewhere in Russia.

His confidence evaporated and was replaced with renewed terror as his cell door opened. Satan. Satan smiled at Ernest's look of fear but kept the rifle at his side and gestured for him to leave the cell. Instead of turning right, towards what he came to think of Nicholas' office, they turned left. They walked up to what looked like a school bus in need of a coat of paint and he was gestured to board. Drab green blankets covered the windows. Satan pushed Ernest into a seat and cuffed his hands to the seat in front of him, binding them tight enough to cause Ernest to wince in pain. Satan got off the bus and another uniformed man approached and placed a canvas hood over Ernest's head.

The drive took less than fifteen minutes. A new guard removed the hood and uncuffed Ernest's hands, almost gently to avoid bruising his wrists. Ernest walked off the bus and caught his first glimpse of what looked like a proper prison, with 16-foot stone walls topped with barbed wire. The steel gate opened, and he walked in. Another guard led him down a dirt path. There appeared to be separate prisons within the larger prison. They entered one of the smaller walled compounds. There were other POWs, ten of them, dressed in pink and gray uniforms and wearing sandals. They looked at him and pointed, talking quietly amongst themselves.

Ernest was led into a room with a dozen cots with personal items below each. The guard left, closing the door behind him. A tall, clean-shaven man wearing the same pink and gray uniform rose from sitting on one of the cots. He was tall but stooped with a definite lean to one side. One of his eyes appeared half shut and a scar ran the length of his face down to his neck. Once the door was closed and they were alone, he spoke. "Commander Wayland Summers, United States Navy."

"Captain Ernest Bellum, United States Air Force," trying to salute crisply but only managing to raise his right arm gingerly.

"We Navy officers don't normally salute without our hats on," Summers said, returning the salute. "But we've adopted your Air Force custom so as to try to get the guards to show some respect. It hasn't worked so far, but one has to try. I suppose you learned this in POW school."

"Yes, sir."

"Arms hurt?"

"A little, sir. I'll be alright."

"Welcome to Hoa Lo, which we call the Hanoi Hilton. This particular compound we call 'Unity.' I am the senior POW in our small group here, the others you saw outside are doing some chores. From your flight suit and wings, I take it you are one of our recent B-52 crews, navigator."

"Yes, sir."

"You guys do good work," Summers said. "I've been a guest of this lovely country for five years now and the way you guys rattled the city, well, it gives us hope. You have the good fortune of having arrived after they stopped torturing us just for fun, and we actually have plenty to eat and proper latrines. But it is still a prison, you can expect rough treatment for not following the rules. About a year ago they put us into these groups where we have the company of others, and we can actually talk. Our group of eleven, well now twelve, are all flyers. You are the only one from a B-52 but we do have three Air Force pilots from other aircraft types. Other than you, our most recent arrival has been in country for more than a year. I think it would cheer everyone up if you could catch us up on the news."

"Yes, sir."

"Word of advice, captain. Those of us who have been here more than a few years have been treated pretty badly. Some of us look like we've been through hell, because we have. You have had the misfortune of getting blown out of the sky but the good sense to have that happen after our conditions have gotten better. Treat those chaps who look well-beaten with some respect."

"Yes, sir, I will."

"Also, try to keep an eye on hygiene, as best you can. Our biggest problem right now is dysentery. It would be unfortunate to survive a SAM and die of diarrhea. Your cot is the one closest to the door, right there," he said, pointing. "Being the junior man, and all that. On top you will find your uniform. The bad guys will want your flight suit and boots. I'll have someone get you acquainted with the latrine and bathing buckets. It isn't much, but it is better than it was."

37. Christmas

Omaha, Nebraska

"Another day without a bomber loss," Lieutenant General Stringfellow said. "That makes three in a row. We did have an airplane hit, however. Purple Zero Two was hit over the Nguyen railroad yard by what appears to have been a 100 mm antiaircraft shell, but the pilot managed to bring the airplane home. His crew, by the way, was hit by a SAM on an earlier mission. They saved both birds."

"Outstanding," General Ham said, studying the results on the slide. "However, we cannot rest on our recent victories. You can be damned sure the North Vietnamese have been watching our change in tactics and will change theirs. How are we going to deal with that?"

"We have studied what we did right and what we did wrong," Lieutenant General Stringfellow said. "On the first four nights we had 300 sorties into Hanoi and lost 11 aircraft, for a loss rate of 3.7 percent. All of the losses came while attacking targets within 10 miles of city center. A disproportionate share of the losses were G-models. So, we will stop sending Gs to Hanoi."

"What's that going to do to our total bomb delivery stats?" General Ham asked.

"Ah," Lieutenant General Stringfellow stammered. "I don't know." He turned from his position at the table and looked to the officers seated in the outer ring of seats. None seemed willing to speak, except one.

"It will not impact our total bombs delivered, general," Major Stephen Bellum said. "Not at all."

"How can that be?" General Ham said. "You can't strip away a quarter of the force with no impact!"

"Actually, general, we can," Stephen said. "It takes four G-models to deliver the same tonnage as one D-model. Since we've capped the top commitment to 30 bombers each night, we have more than enough D-models to cover the total commitment."

"Then why in the hell have we been using the G-models at all?" General Ham asked.

"I don't know," Stephen said.

"Is that all?" General Ham asked.

"Yes, general," Lieutenant General Stringfellow said.

"I wasn't asking you," General Ham said. "Major?"

"Yes, general," Stephen said. "The word from the EWs in the field is that whenever you have less than three aircraft jamming, the odds of survival go way down."

"Is that true?" General Ham asked, this time directing his question to Brigadier General Mayfield, the Director of the Electronic Warfare Division.

Mayfield stood, uneasily. "We have been studying that possibility for quite some time, general," he said. "We don't have enough information to make that assessment."

"Is that true?" General Ham asked, looking directly at Stephen. The room became utterly silent. Stephen realized immediately that he was about to have what many called a "career defining moment" by contradicting a general officer in public. But the general officer in question was taking credit for actions in the field that were contrary to his earlier orders. There was only one thing to do.

"When I passed the information to the Operations Staff, I was told there was no data," Stephen said. "So, I looked at the losses we experienced at that point and found there were more than a few losses tied to instances where a bomber had to drop out and a three-ship cell became a two-ship cell. We also lost airplanes when one or more of the bombers had less than a full complement of jammers, decreasing total jamming potential of the cell and increasing the ability of the enemy radar to burn through the jamming."

"Did you get this from your brother?" General Ham asked.

"No," Stephen said, weakly. Colonel Snedeker rushed to General Ham's side and whispered. The staff fell silent and Stephen felt that every eye of the SAC Operations Stafff was on him.

"Sorry, major," General Ham said. "I didn't know."

The staff sat quietly with some whispering. Stephen sat down.

"As you all know, the North Vietnamese walked out of the peace talks," General Ham said. "The President has ordered a massive attack on Hanoi tomorrow."

"Yes!" General Stringfellow said, pumping his fist.

"I have placed a call to the commander of the 8th Air Force," General Ham continued. "I am delegating all planning to them." The staff sat in stunned silence. "They are to take all the lessons learned and choose the tactics they see fit. I don't want to hear of anyone in this headquarters staff meddling in their affairs."

38. Linebacker II, Night 8

Hanoi, North Vietnam

As he had been warned, many of Ernest's fellow prisoners looked pretty well-beaten but happy to hear of any news he had about life back home, especially about the election and its aftermath. He didn't know anyone in his group and guessed he was the youngest in terms of age and rank. After nearly a week in his flight suit without a bath, he was grateful to be handed a bucket and a bar of what was supposed to be soap. Their group was allowed to shave every other day and that too was therapeutic for the soul.

Without uniforms Ernest had to associate faces with names, branch of service, and rank. He thought he would have a kinship with his Air Force peers, but they didn't seem interested in his company. They were the three in the group who seemed to have spent the most time in hell, so he didn't take their behavior personally. He was, he realized, now more aware of other people's feelings than ever before.

"Navigator?" the POW shaving next to him asked.

"Yes," Ernest said.

"Me too," the now clean-shaven prisoner said. "Dan McDaniels, backseater in Navy Phantoms. I was the junior man here until you showed up."

"We got hit by an SA-2 dropping iron bombs on a railroad station," McDaniels continued. "Our ECM gear is primitive compared to yours. When we got the Fan Song, there wasn't much I could do other than hope the pilot could out-maneuver it. He didn't. It was the longest thirty seconds of my life. I didn't think an SA-2 could touch a Buff."

"I felt the same way a few months ago," Ernest said. "But I learned the hard way we are not so untouchable. I'm still not sure how they did it."

After his shave Ernest paced a line along the perimeter of their compound. Many of the prisoners favored pacing the smaller courtyard and two would bring their cots outdoors to sunbathe. The entire concept of sunbathing in the Hanoi Hilton struck Ernest as funny. But he kept that thought to himself. He heard that many of the old timers could at first only communicate by tapping codes through walls, until the bad guys figured that out. The new prison cells were constructed to prevent that. Now, years later, they finally had the chance to talk to compatriots, but few chose to do so. For the first time in a long time, Ernest felt loneliness.

He was haunted by thoughts of *her* and the last time he saw *her*. She had given another detailed and well delivered weather briefing to his crew and those of the other aircraft in his cell formation. He felt a little anger at how some of the pilots from the other crews tended to flirt with her, and perhaps personally wounded whenever she smiled at another male. But once everyone started congregating near the exits, he had rushed back to see her.

"See you tomorrow?"

"Of course," she had said. It was a bit noisy, and he didn't hear her words precisely, so he leaned forward. And before he knew it, she planted her luscious lips right on his. It was more than sensual, it was delicious. But then he heard his name.

"Duty calls," he then said. She smiled and winked at him.

That was the last thing he said to her and it could be the last thing she will have ever heard from him. *Duty calls?* If only he had said something else. Anything else. Regret was not something Ernest felt often, but he felt it now. He longed for the taste of her.

But he still had problems to solve. Ernest mostly thought about his last ten minutes in the air. As an EW – he had yet to admit that to any of his fellow POWs – he had a greater understanding of the SA-2 than most American aviators.

Something didn't add up. "It was the longest thirty seconds of my life," McDaniels had said. That half-minute was pretty typical in training. The Fan Song would identify you, triangulate you, launch the missile, and then it was the intense game of matching jammers to frequencies over and over again.

"I am smarter than this!" Ernest reminded himself. He had already figured out the problem with jamming and had hoped to pass that on to his brother, but it was too late now. Ernest figured this out during the last thirty seconds before life went to hell for him.

"Thirty seconds," Kevin had said at the time. "Doors. We are on course."

"Uplink," was Ernest's next word. "One ring. Fangs."

"Lead's in the turn," Harry said next, followed by, "time to drop?" At that moment, the fangs on Ernest's scope not only grew in height, but they were accompanied by two more sets of fangs. They had multiple Fan Songs locking in on them.

"Twenty seconds. Stop maneuver, level," were the next things Kevin said.

"Launch!" Ernest said next. "Three and a half rings, headed right for us."

The last thing he remembered hearing from Kevin was "not yet, five more seconds."

But then he rewound the recorder in his head to Harry. "Lead's in the turn."

And that was half of the answer Ernest now realized. The combined effect of the three aircraft in the cell worked until the cell was no longer a cell. Everything changed with the post target turn. When lead wrapped into a sixty-degree bank right turn, his jamming antennas were no longer pointed down to the SAM, they were pointed sixty-degrees to nowhere. The cell was down a third of its jammers. When number two turned, there went another third and number three became solo. One B-52's jammers could not emit enough noise on the uplink frequency to jam the missile.

He had flown the same bomb run three times, the first and second times as number two, the last as number three. When the cell turned to the target and changed from a triangular formation to a single-file line, the jamming signals from the cell got weaker, that made sense. Number two was in the center of the cell. Of course, it made sense. But as number three, it got even worse at the IP, following the turn south to the target. And now, sitting in his shared room at the Hanoi Hilton, that made sense too. If you are on the left side of the formation and the formation turns right, you will have some catching up to do. And the further behind you fall, the less protection you get from the other two aircraft. He had to get this information to SAC.

"Anybody ever escape?" he asked Commander Summers.

"Only in the movies."

* * *

Ernest prepared himself for another night of bombing, but the old timers assured him the U.S. always gave everyone Christmas off. Sure enough, it was a quiet night with members of his compound reciting from newly acquired bibles. Ernest was raised Catholic, but members of his family called themselves "part-time Catholics." Ernest declined the chance to take turn in the readings, but he sat quietly and continued to think of the math problem that was eluding him. He now understood why the jamming didn't appear to work at the most important time. But why didn't he have enough warning time?

The night after Christmas was not quiet at all. At first there was a single explosion with a three second pause. And then it was as if hell itself was erupting from the center of Hanoi with no time to rest between explosions. The secondary explosions very near to Camp Unity provided light and Ernest could see his fellow POWs were not at all scared, as he realized he was. They were energized. A few gave him a look and a thumbs up. "That Buff sure can pack a punch!" Ernest feared the next bomb was headed off course, right on top of them.

"Not yet," was one of the things Kevin said just before he, well maybe he didn't die. But Kevin wasn't going to drop his bombs anywhere but their assigned target. Ernest didn't feel the same fear anymore.

Soon after it started, it ended. He kept waiting for more, but more never came. "Concentrate your forces," Harry had said. Maybe SAC had finally figured that out.

39. 15 Minutes

Omaha, Nebraska

"It was 120 B-52s in one massive strike coming in from six directions and leaving in seven different directions," Lieutenant General Stringfellow said while pointing to a map of the Hanoi area drawn on the now familiar butcher paper on easels. "And all of it took place in just 15 minutes."

General Ham sat back in his chair, stoically. Stephen sat, thinking that Stringfellow was practically taking credit for the decisions made half a world away. The decisions that overruled everything the SAC Operations Staff had originally planned.

The staff sat back in awe of the wonderfully drawn map with hand-drawn bombers coming in from what looked like random directions and leaving in others. General Ham appeared to be deep in thought as he studied the maps, giving the staff time to talk quietly about how "we" did such a masterful job.

"All indications are that we completely overwhelmed North Vietnamese defenses," Stringfellow continued. We dropped over 2,000 tons of bombs." The word "we" started to grate on Stephen.

"There were, unfortunately, two losses," Stringfellow continued. "Ebony Zero Two was hit by a SAM but we did see at least four chutes. Ash Zero Two was hit but made it to Thailand where it crash-landed. The entire crew made it out okay."

"Was it the post target turn?" General Ham asked.

"No sir," Stringfellow said. "Most of the bombers simply flew in one way and out the other, using very small turns. The biggest turn was still less than 90 degrees, and all those aircraft made it out okay."

"Then what happened to Ebony Zero Two and Ash Zero Two?" General Ham asked.

"We are still assessing that," Stringfellow said.

"What do you have to say about that?" General Ham asked, looking straight at Stephen.

Stephen rose to his feet and noticed Stringfellow's intense glare. A number of the general officers made a point of looking at him, wide-eyed, unblinking, willing him to remain silent. Stephen made a quick decision that he knew would impact his career forever.

"General I think there is a lesson from both of last night's losses that should immediately change our ROE if we hope to avoid a repeat." He had everyone's attention. "In both cases, the aircraft were in cells where one of the three aircraft aborted either on the ground or en route. Our ROE tells them to press on with one less aircraft. But that reduces their jamming power by a third and, as we've seen too many times, that's all a good SAM radar operator needs to burn through."

General Ham looked to his top Electronic Warfare Division general. "What have you got to say?"

"I agree with the major's assessment," Brigadier General Mayfield said.

"Good job Major," General Ham said. "I wish half the officers in this room had your kind of guts." He looked at Stringfellow. "Make the change, immediately."

Stephen sat, knowing that he had won the battle for the day, but would lose his career. The Air Force didn't look kindly on officers willing to put a knife in any general officer's back. As the staff continued to discuss what needed to be done to change the ROE, Stephen reflected on what he had done and why. "Everything happens for a reason," his dad would say whenever some kind of setback befell his boys. And he was usually right.

One of General Ham's orderlies came in with a printed message. General Ham read and smiled.

"I have some news that may interest you," General Ham said, raising his voice. "I just got word from the White House that the

North Vietnamese government has contacted the United States government, offering to restart Paris negotiations on January 8th. President Nixon told them January 2nd. And he said this to us: 'Keep bombing.'"

40. Epiphany

Hanoi, North Vietnam

"Skipper's gone to get our mail," McDaniels said. "They allowed each of us to write a letter every month about a year ago. And we started getting mail about six months ago. It is downright civilized."

When Commander Summers returned, he had a letter for everyone except Ernest, and a big package for himself. He asked for some privacy as he opened his gift, so everyone vacated to the courtyard.

"I bet it's a cake," McDaniels said.

"With a file in it?" someone else offered.

Summers came out into the open. "Gentlemen," he began. "As most of you know, and some of you can verify, I spent the first three years here in solitary confinement with nothing more than the clothes I brought with me when I arrived. Conditions have improved. And now I can say they have greatly improved." He reached back into the open door. "Now we have a basketball!"

The group cheered and drew out a makeshift court in the dirt and spent the rest of the day inventing a hoopless game of basketball.

As they played, Ernest drew triangles in the dirt. He drew the top view, "God's view" in aviator speak, and the side view of an SA-2 operator's problem. Looking at the former, he knew the operator needed to solve azimuth. You have to know where to aim, left to right. Looking at the latter, the operator had the same problem in the vertical, something called elevation. You have to know how high or low to aim. What's missing?

"Range," he heard from an approaching POW. He looked up to see McDaniels.

"Range?" Ernest asked.

"You are looking at SAM firing solutions," McDaniels said. "It is a three-sided problem: azimuth, elevation, and range. You not only need to aim the thing in the right part of the sky, you have to fuse it so that it blows up at the right time."

"I thought it was a proximity fuse," Ernest said.

"It is. But they want it to blow up after a certain distance, in case they aren't close enough for the proximity fuse to go off. I think that's what got us. We were doing a pretty good job of jinking away, but when the missile went off, the shrapnel turned our Phantom into a glider on fire."

"Which is why they need the height finder," Ernest said. "The Fan Song gets the azimuth and the elevation, so they know where to shoot, but they need the height finder to triangulate the distance."

Ernest looked at his drawings anew. Knowing the height of the target allowed the Fan Song fire-control operator to determine the distance to the target. A-squared plus B-squared equals C-squared. A was the height, B was the distance over the ground, a C became the distance to target.

"I think that's how it works. We had airplanes on the boat that did nothing but target the RSN-75 Fan Song. You take out the Fan Song, the SAM can't get you. It isn't only for missile guidance, you know. They need it for target acquisition too."

"I've been thinking about this backwards. I have been wondering how they can get the Fan Song to guide the missile without aiming it first. That's the wrong question. The question is: how can they aim the missile before activating the radar? If they can somehow aim it first, the guidance part is easy.

"You just aim it using something else. But that's hard to do with enough accuracy. The Fan Song beam width is pretty narrow."

"Seven-point-five by two-point-five degrees."

"I'll take your word for it," McDaniels said walking away, suddenly more interested in playing an invented game with a basketball.

"SAC Headquarters," Ernest said aloud, realizing the answer at last. "If you know the route is going to be the same, all you need is the time at some point along the route to solve for azimuth, elevation, and time. And since the altitude never changes, you also have height. "We are making it easy for them."

It was another quiet night in Hanoi.

* * *

The next morning, Commander Summers returned from his daily meeting with the other senior prisoners and called for Ernest.

"I got news for you," he said. "Both of your pilots are guests in the next compound." Ernest's face lit up, but he couldn't form any words. "Your pilot got hurt when he landed, we think his leg is broken. Your copilot is as fit as a fiddle."

"Can I see them?" Ernest finally asked.

"They don't allow prisoners from one compound to mingle with the prisoners of other compounds, but there is a loophole. They usually have two prisoners take the daily food rations on a cart from compound to compound. This is usually choice duty, since it gets you a chance to look around. I pulled some strings and got you assigned for today's rounds with Captain Broadmore."

"When?"

"Right now, present yourself to the guard at our gate and say 'food duty' and they'll take you there. They'll have a cart made up for you. You just push it to each gate and a prisoner from that compound will tell you how many they got, and you give them that number of packages. Keep your voices down and try not to look like you are passing along state secrets to your pilot. Get to it."

Ernest presented himself to the guard, the gate was opened, and he was led to a central building. At the door he saw Captain Harry Broadmore dressed in the same pajamas and sandals, loading food packages onto a cart. "Harry," he said, failing to hide his grin.

"Ernest," Harry said with the same poorly disguised smile. "You want to push or do the handing out?"

"I can push." They followed the path, led by a guard, and stopped at the first compound. Harry approached a prisoner at each gate, exchanged a few words, and handed over the requisite number of packages.

"How many?" Ernest asked as he pushed the cart to the next compound.

"You, me, and the boss."

"That's it?"

"Yeah."

"How can you be sure?"

"I heard your seat fire and looked at Major Stevens. He nodded his head and yelled what I assumed was 'eject, eject, eject.' I didn't hang around after the first 'eject' and I punched. I kept my eyes opened the whole time. I saw your chute and I saw the nose of the airplane pitched over. It was detached from the rest of the airplane pointed almost straight down but the wings and the rest of the fuselage were a big ball of flames. I saw Kevin and Tommy eject right into the fireball. They didn't have a chance."

"Polanshek?"

"I didn't see any chutes from back there, but it was a big ball of flame."

They arrived at the next compound and Harry busied himself with handing out packages. Ernest replayed the last moments in his head. He didn't remember the 'eject, eject, eject' call. As the closest crewmember to the bomb bay, Ernest had the least chance of surviving a direct hit there. And yet he survived. The navigator and radar navigator ejection seats fired down instead of up. "They didn't have a chance," Ernest repeated.

"Did you get beaten?" Harry asked while walking to the next compound.

"A little, but it seems they lost interest in me after a few days. They tried to get me to admit to being the EW, but I put on the navigator act." Harry laughed. "What?"

"Major Stevens was thinking of you the minute we met up," Harry said on the way to the next compound. "We found each other before we got captured and I told him about the navigators. Stevens told me that our story would be that Tommy was the EW and you were the navigator."

"Oh," Ernest said. "Tommy died and that made it easier for me to live."

"The boss ordered me to not divulge our names but if it got too unbearable to remember you were our nav and Tommy was the EW. They didn't really mess with me, but Major Stevens got beat up pretty bad. His face shows it. But he said once he gave them the name of our EW, they backed off." He smiled that Broadmore smile.

"It was as if we had it all planned," Ernest said.

"Yeah, and we owe it all to the way the pencil pushers got mad at you for jumbling the order on our flight logs. Everything happens for a reason, I guess."

"It does," Ernest said. "Sometimes the reason is hidden, but there is a reason."

41. Gray 03

Omaha, Nebraska

For the first time since Operation Linebacker II had begun, General Ham let it be known he would be the only speaker at the daily SAC Operations Staff meeting. Stephen entered the room to find every seat other than those reserved at the table had already been taken. There were generals he didn't recognize seated in the outer rows. A team of photographers stood ready nearby.

"Standing room only?" Colonel Snedeker joined Stephen at the perimeter.

"Yes, sir. Any ideas?"

"It's all good, Stephen. I just left CINCSAC's office. He's been on the phone, but he'll be here soon. It's all good."

One-by-one, generals of the SAC Operations Staff entered and took their places at the table, but only after shaking a few hands along the way. Stephen had to remind himself that good news for SAC was in no way a personal attack on his fallen brother. He had to regain whatever objectivity he had before Linebacker II had begun.

Lieutenant General Stringfellow entered and shot the generals at the table a hearty thumbs-up. The seated generals began to clap and those in the outer orbit of seats followed suit. General Stringfellow smiled but extended both hands, palms down, in a tamping motion. "Boss is on his way," he explained.

Stringfellow sat down and turned his head to the entryway, a sign to for the staff with seats to remain seated, and for everyone to be quiet and be ready to spring to attention at the right instance.

Thirty seconds later, as if choreographed, General Ham entered. Everyone stood at attention, eyes caged forward as if military cadets in school. General Ham walked to his position at the table and extended a hand. "Alan, helluva job"

"Thank you, general."

"Everyone, be seated," General Ham said. Once the shuffling of seats subsided and all eyes faced the four-star, he smiled and spoke.

"The North Vietnamese have agreed to our January 2nd peace negotiations in Paris," he began. "President Nixon has flashed a message to all US bases that Linebacker II will cease today. At 2345 North Vietnam time, Gray Zero Three, the last B-52 in the last wave scheduled for today, dropped its bomb load on Trai Ca and returned safely to base. Linebacker II is over."

42. Homecoming

Omaha, Nebraska

It was a typically cold and windy Nebraska plains day and Stephen, for a brief moment, thought of calling in sick. The ceasefire with North Vietnam was signed on January 27th and included provisions to release the nearly 600 prisoners of war within 60 days. Since his brother was officially killed in action, it would be a bittersweet celebration. His thoughts of late had turned to the airlines but getting a job as an airline pilot would be tough with all the transport pilots being released from active duty after the war. But duty, as they say, is duty. He donned his uniform and made the drive to the headquarters building.

The roads had become sloppy after an unseasonable warm up to the fifties was followed by a dip back below freezing. The temperature couldn't make up its mind in the last week and the precipitation mixed between snow and rain. The result was slop. Stephen left his car, two months since its last wash, and negotiated the equally filthy parking lot to Building 500. He scrubbed the bottom of his shoes on the inverted brushes near the entrance and entered the heated building still wearing his heavy coat.

He entered the offices of CINCSAC's Action Group where all his officemates were already at their desks, busy little bees with too much to do, and too little time. Colonel Snedeker was in the "make work" mode of operation, and their office tempo had gone up, not down, since Linebacker II had concluded. But Stephen had somehow escaped the extra work, perhaps still off limits to his boss because of, well, because.

His inbox was filled with the normal ebb and flow of staff work, but was topped by the morning "Early Bird," a collection of news stories faxed from the Defense Department and required reading by all staff officers in the headquarters staff. It was nothing more than articles from various newspapers cut and pasted into a fax, but it gave a good summary of what those calling the shots in the U.S. military needed to know.

The Early Bird had lately been filled with Air Force plans for "Operation Homecoming," a plan to finally repatriate all known Prisoners of War held in North Vietnam. It was exciting news for most in the Air Force, indeed for the entire country. But it was news that Stephen was unable to hear. While it bordered on heresy in his office, Stephen added his initials to the Early Bird and moved it to his outbox, unread.

Next in line was an issue paper about the need to close bases and reshuffle some of the bombers returning from Southeast Asia. It would take the rest of his morning to plow through the report, but it would cut through the boredom. It was important work and Stephen suspected he would lead the office effort.

This is my life now, Stephen thought. Even as a staff officer, he never failed to summon the enthusiasm for whatever it was he was doing. The drive to do something he cared about it was drove him to working harder than his peers. It was a drive he feared he would never have again.

"Stephen," Colonel Snedeker said, peering from his inner office. "General Ham wants to see us both, right now."

"Just like the old days," Stephen thought. "Any ideas, sir?"

"Not a clue," Snedeker said. "But his secretary made it clear your presence is needed."

The secretary gave them her customary, "just a moment please." She popped into the office and came back immediately. "CINCSAC will see you now."

Colonel Snedeker led the way but stopped short, took a step to one side and allowed Stephen to come up alongside. General Ham stood, flanked by the top generals of the SAC Operations

Staff. Only General Ham smiled, the others appeared stoic at best.

"Stephen," General Ham said warmly. "Why don't you have a seat with me." He gestured to two seats pulled up to a small table. Stephen sat, his mind racing.

"I wanted to be the one to tell you," General Ham said. "We just got the official list of all the POWs about to be released. Including Captain Ernest Bellum."

43. The Proposal

U-Tapao, Thailand

Second Lieutenant Marjorie Ralston was halfway through her shift, having completed all the crew briefings scheduled for the day. The last departure was at 1400 and she anticipated another day marking time until she too would be allowed to leave. Her boss, a captain, appeared at the door.

"I am out of here, goodbye U-Tapao!" he said. "You okay?"

"Yes, sir," she said. "Just another week of this?"

"That's what they say. They just announced the wing will officially deactivate on Thursday. The last B-52 is already gone. If you don't catch a ride on that last tanker next week, you become a local."

"I can think of worse places," she said. "But I won't miss that flight."

"Don't forget to get your permanent gear onto our shop pallet by tomorrow. They aren't going to allow you to bring more than your B-4 bag and one personal item on the tanker."

"I already did, sir. I have only what I need now and a violin."

"Violin? Are you sure they will let you take that?"

"I checked with the tanker squadron and they said if it's my personal item, it is okay. I couldn't bear to lose it. It is my most important possession."

"His?"

"Yes, sir."

"I'm sorry." The captain lingered just a little longer, his eyes on the floor. "Oh yes, I almost forgot. I got a call from the wing commander. He's going to drop by shortly. He wanted to make sure you are here."

"The wing commander? Me? What's it about?"

"He didn't say. Probably just a pep talk to the people scheduled to leave on the last airplane. I'm sure it's nothing to worry about. Okay, see you back in the states."

Nothing to worry about? That was easy for the captain to say! But having a visit from the colonel himself was certainly something to worry about! The weather shop was neat only because it was empty, most their gear already packed away. She thought she should find a mirror to make sure her uniform was in order but it was too late. Colonel Matthew Stipp appeared at the door and walked in. She came to attention.

"Lieutenant Ralston?"

"Yes, sir."

"As you were," he said. "Please sit down. In fact, I'll sit down too."

"Is there are problem, sir?" she asked while sitting.

"No, in fact just the opposite. We just got the word of the final POW list. Captain Ernest Bellum is alive."

"He's what?" she cried.

"We didn't know until this morning. He and both his pilots managed to eject. The North Vietnamese hadn't released any of their names until this morning."

"I, I . . ." she started but couldn't continue, her sobs overtaking any speech.

"I had the same reaction when I heard, believe me!"

"Can I speak to him?"

"Not yet. But maybe soon. I just got a call from the Air Attaché in Saigon, Major Jack Haverill. He was an academy classmate of Ernest's older brother. He just got back from Hanoi last night where he got to inspect several of the POW camps, and he met Ernest. Said he looks great. He got hurt in his arms and shoulders, but he's otherwise healthy."

"I am so happy to hear that. I can't wait to see him."

"Major Haverill is going to call you here in about an hour. He says he has a message from Ernest. So don't go anywhere for an hour." The colonel got up.

"I won't! Thank you!" she said, standing. She suddenly lunged at him, hugging him. "Thank you!"

She pulled away, blushing. "I've never hugged a colonel before, sorry sir."

"Don't worry about it," he said. "I wish all the news I had to give around here was as happy as this."

After the colonel left, she couldn't sit. She rushed to the office latrine to look at her face, touch up what little makeup the Air Force permitted, and re-bundled her hair into the neat bun required of women officers with shoulder length or longer hair. When satisfied, she let out a laugh. She had fussed about her appearance for a phone call. She didn't have to wait long. The phone rang.

"Lieutenant Ralston."

"Hi, I'm Major Haverill from the Saigon Embassy, I trust Colonel Stipp has already told you about Ernest?"

"Yes, sir, it is wonderful news."

"It is that. I never met Ernest before but the minute I saw him I thought he was a younger, better-looking version of Stephen. You can tell Stephen I said that."

Marjorie laughed. "I'll leave that to you, sir."

"He asked me to call you with a message as soon as possible, I hope you are ready for it."

"Yes, sir, please, go ahead."

"Okay, here goes. Message begins. Ernest says, 'Marjorie, I love you. Will you marry me?' End of message."

Marjorie looked at the phone for a second and then erupted. "Yes, yes, yes! Please tell Ernest yes. And that I love him dearly."

"Well, I'm glad that turned out that way!" Major Haverill said.

"When can I see him? When can I talk to him?"

"Those details are still in the air. We think they will be released in order of their dates of capture. So, I am afraid Ernest will be one of the last to be released."

"Of course, that is only fair."

"We think it will take two to three months, so he could be released in April, or maybe as early as March."

"That's wonderful."

"They don't have any way of allowing them to make phone calls, but they will be allowed mail service. I'll get you the address."

"Sir? When will you see Ernest next?"

"I have three other camps to visit, but I hope to see him next week."

"Oh dear, we are closing the operation here at U-T next week."

"Don't worry, Marjorie. I'll pass along your answer. I'll also have someone here get your contact information and pass along mine. You're getting married after all."

44. April 4, 1973

Omaha, Nebraska

It was her, they all realized. They were gathered in the VIP lounge at the Offutt Air Force Base passenger terminal. She was dressed in the standard Air Force blue uniform with a dark blue skirt topped with a dark blue jacket, first lieutenant bars, a row of ribbons, and a nametag that said Ralston. She closed the door behind her. Stephen was the first to approach her.

"You must be Marjorie," he said.

"Yes, sir. You must be Stephen. Ernest has told me so much about you."

"I hope not! Let me introduce you to your new family."

The family was all in attendance. She addressed Jonathan as "Colonel."

"I'll have none of that!" he said. "You call me dad. And that's an order!"

"Yes, dad, of course."

"Marjorie, I'm Mary, Ernest's mom."

Marjorie extended a hand, Mary extended both and wrapped them around her. "In this family the men shake hands, the women hug."

"That's wonderful," Marjorie said.

"As you can imagine, we have a bit of pomp and circumstance happening," Stephen said. "Ernest is only the second POW arrival for the base here. The news media is outside, and we've let them know they can take video of the airplane arrival, Ernest getting off the airplane, and you rushing up to see him. But no interview. The family will stay here. Then you and Ernest walk the red carpet into this lounge, no press, and we can have a real reunion."

"That sounds so wonderful," she said. "I feel a bit guilty getting to see him first."

"Don't," Stephen said. "I talked to Ernest yesterday and he insisted on this. This is his day, so he gets his wish. For today, anyway."

"Where are you two going to live?" Mary asked.

"The Air Force is giving Ernest a choice of an electrical engineering job near Boston, becoming an instructor at Keesler, or going to the Academy to teach. He hasn't decided yet."

"What about you?"

"I've not seen Ernest since before he was shot down and we've only talked on the phone a few times since. We need to sort these things out. The Air Force is offering me early separation if I want it. I want to be wherever Ernest is. So, if they can't station us together, I may leave the Air Force. My priority is Ernest."

Mary gave Marjorie another hug. "I'm so happy to hear that." Susan approached.

"Hi," she said, "I'm Susan. Stephen's wife. The young lady and the two boys over there are ours: Jennifer, Jeffrey and Justin."

Marjorie extended her hand. Susan extended both arms and hugged her instead. "Remember, in this family the men shake hands. We women hug."

"That is how I was raised too," Marjorie said. "I have so much emotion in me I might burst. In fact, I need to touch up my makeup. Ernest will think I'm a mess!"

The women went off to the ladies' room while Stephen and Martin stood, looking out to the tarmac from the glass windows. "You should introduce yourself, Martin," Stephen said.

"I will. I just wanted everyone else to get their chance first."

"Graduation is coming up, isn't it?" Stephen asked.

"June. You coming?"

"Wouldn't miss it. What's after that?"

"Not the army, thankfully. I was sweating it out until they cancelled the draft call up in January. The college accepted me for the master's program. They are going to pay for it, in fact."

"That's great!"

"I knew you were going to be a history professor!" Jonathan said.

"Not yet, I still have a long way to go."

The three looked out the window, seeing a cargo aircraft taxi by. "The war was a waste, wasn't it?" Martin asked.

"Let's keep that kind of talk to a minimum today, Martin. Okay?"

"Sure," Martin said. "We owe Ernest his day. I'm very happy for him. For them, I mean."

"You should say something to her."

"I will."

The women returned and gathered around a coffee pot. The passenger terminal provided a basket of fruit, a plate of cookies, and coffee. Everyone helped themselves to the coffee, but the food went untouched. After a few minutes, Marjorie made her way to Martin.

"You must be Martin," she said.

"Must I?" he said. She laughed.

"Ernest says you are going to be a history professor, says you have an encyclopedic mind for that kind of thing."

"He did?"

"Yes, of course. We once talked about General Patton's fall from grace before and after D-Day. He said I should pick your brain about the parts of Patton's life not in the movie."

"The stories I could tell you!" he said. "Did you know . . ."

"The plane just landed!" Stephen said, interrupting all conversation in the lounge. "You can see it on the runway from here."

Everyone gathered at the window. A colonel showed up at the door. "Where is Lieutenant Ralston?"

She approached the colonel. "Here, sir."

"I'm Colonel Tim Bradley," he said, "the 55th Wing Commander. I will be escorting you down the carpet. We need to take our positions now."

She smiled at the Bellum family and followed the colonel out the ramp side door to the flight line. He took his position on the red carpet with Marjorie to his left. As the airplane came into view her face lit up. The C-9 Nightingale was one of the smallest versions of the McDonnell Douglass DC-9, made all the more distinctive by its white fuselage and red cross on the tail. The aircraft was very noisy, almost as noisy as a B-52, but the colonel and Marjorie remained at attention. Several reporters stood in a roped off area with a view of the aircraft, the red carpet, and the two waiting officers. Those not holding cameras held their ears. Once the engines were shut down, the sounds of camera shutters filled the air. The door at the front of the airplane opened and an automatic stairway appeared below the door and unfolded itself. A sergeant dressed in a green flight suit got off the airplane and stood at attention beside the stairs. Ernest appeared at the door.

He was dressed in his Class A uniform, all blue, captain's bars on his shoulders, and four rows of ribbons on his chest. He put his flight cap on, spotted Marjorie, and ignited his best smile. As soon as his foot hit the red carpet, Marjorie could stand it no longer and ran to him.

She hit him with so much speed she nearly knocked him off his feet. He hugged her tightly and they kissed, repeatedly. "This has to be a dream," she said.

"Let's share all our dreams," he said. "From now on."

"That is just what I wanted to hear." She released him, wiped the lipstick off his mouth with her hand, and they walked down the red carpet. Ernest saluted the colonel. The motion still lacked the military snap required of Air Force officers, but Ernest's grimace telegraphed the pain the movement was causing.

"Welcome home, captain," he said.

"Good to be home, sir."

The colonel pointed down the red carpet and led the way. He opened the door to the VIP lounge and allowed the couple to enter. Once inside, the colonel paid his respects to retired Colonel Bellum and Mary, and then left.

After the ladies released Ernest from their hugs, Stephen shook Ernest's hand. "Welcome home brother, sorry we messed up the war so much."

"I hear you fixed things on Day Five," Ernest said. "Better late than never. Besides, I think it made it easier for us at the Hanoi Hilton. You really had them scared."

"I hope so," Stephen said.

"No more bombing campaigns in your future?"

"No thankfully. They got me attached to review the Linebacker Two after-action reports from here all the way down to the wings. We need to document how badly we screwed up so we can learn from our mistakes."

"Sounds very un-SAC like."

"Well, that is my mission. Your crewmates cannot have died in vain. I am sorry to hear about the navs."

"Me too. Did we ever get anything official on our gunner?"

"They changed his status to MIA."

Martin approached and extended a hand, which Ernest shook. "It felt like the world came to an end when we thought you were dead, Ernest. I was very happy when the news finally came out that you were okay."

"Thanks, Martin. I was happy when I heard about the draft call up being canceled. Everything happens for a reason, doesn't it?"

"It seems to." Martin caught his father's eye and stepped aside. Ernest looked at Jonathan and extended his hand.

"To hell with that!" Jonathan said, pulling his son to him, hugging him tightly, tears streaming from both men.

"They put you through hell, son. And I feel like we've been there with you."

"I know, dad," Ernest said. "It is all over now. We have our lives back."

"The gift of life must be appreciated, to be lived to its fullest, Ernest. You know that better than any of us now. You embrace that. You and Marjorie, embrace that."

45. The Lab

Lexington, Massachusetts

Major Wesley T. Boyle returned the crisp salute offered by the guard at the Hanscom Air Force Base's main gate and gave the sergeant a polite, "good morning." He recognized the salutes were quicker and the "sir" offered by the enlisted troops were more respectful now that he had the gold oak leaves of his new rank on each lapel. He pulled into his parking spot, the one marked "AFSC/ESD/D-2" which was Air Force speak for him and him alone. Before exiting his car, he flipped down the sun visor to check the mirror he had glued into place. His hair was neatly combed and the five ribbons on his light blue shirt were exactly centered. In a few days he would add a sixth ribbon for an even two rows of three ribbons each. It was an order to things all engineers could appreciate.

"Good morning, sir!" rang out multiple times. Major Boyle was happy to see each member of his staff were already at their desks. The military members each dressed in their light blue shirts and dark blue pants or skirts. Each were wearing their ribbons proudly, as he had directed when he took over, a month ago.

"Good morning, everyone," he said. "Another beautiful Massachusetts day."

Major Boyle entered his private office, leaving the door open behind him, just as suggested by the book of "Effective Leadership Tactics" he was issued before taking over the D-2 branch of the Electronic Systems Division of the Air Force Systems Command. It was almost ten years to the day since he became a commissioned officer in the United States Air Force, and he had made major at the nine-year point and was able to pin on the rank a year later.

"Not bad," he said to himself whenever looking at his newly acquired rank. He checked his image in the mirror he had stashed in his top desk drawer. Everything was still as it should be, except the ribbon he was about to get in a ceremony on Tuesday.

Staff Sergeant Ellis, the office secretary, entered and placed a stack of files onto his desk. "MSM," she said. "Not too shabby, sir."

"Yes, I guess they give those out now and then," he said. He took the top folder from the stack and saw the script for the awards ceremony was in order. It would be the ESD colonel presenting awards to four officers in total, a lieutenant colonel and three majors. Only he and the lieutenant colonel would be getting the Meritorious Service Medal. The two other majors would be getting the less prestigious Commendation Medal. It would be a good ceremony for the D-2 branch, seeing their new boss with an MSM awarded by the colonel. The entire ESD staff would be present, nearly three hundred officers, enlisted personnel, and civilians.

The second folder in the stack came from the personnel division and in it he found a standard computer printout on eighteen-inch-wide paper with alternating white and light green stripes. The pages were perforated with holes on the left and right margins as needed by the printers. He carefully folded each side and tore off the edges. He found the holes in the borders distracting and trained himself to ignore the printed matter until the page was presentable. It was a standard personnel flow sheet, showing gains and losses to the D-2 branch. Staff Sergeant Ellis had circled one line in particular and wrote, "Today!"

"A new captain!" Major Boyle said to himself. His branch of twelve would become thirteen. The printout only gave the captain's name, Air Force Specialty Code, year group, and education status. Captain Ernest Bellum had an AFSC starting with the characters 12B, was in the 1969 year group, and was an Electrical Engineer from the Massachusetts Institute of Technology. It was mostly good: the Captain was four years junior to him, so he wouldn't be a political threat. Having an EE on staff was good too. The 12B AFSC didn't make any sense. All the engineers on staff had codes starting with 3E5.

He picked up the phone. "Sergeant Ellis, what is a 12B AFSC?"

"I'll find out, sir," she said.

He knew the "1" meant the officer was an operator, such as a pilot or a navigator. But mere operators didn't have the necessary security clearances to work at ESD. The D-2 branch was highly classified, dealing with Electronic Counter Measures. No mere pilot or navigator could possibly have the necessary clearance.

"B-52 bomber electronic warfare officer," Staff Sergeant Ellis said from the door.

"It must be a mistake."

"I asked, sir. He has all the right clearances, and his assignment was okayed by General Sams himself."

"General Sams?" Major Boyle sat back in his chair and started to think his neatly designed world might be in for some trouble. If the general in command of the Air Force System Command wanted a mere captain placed in the D-2 branch, the mere captain had friends in very high places. "Any word on when?"

"Today, sir. He's at personnel right now and they say he'll be released to us at 0900."

At 0905, Major Boyle picked up the phone. "Where is he?"

"I'll find out." Boyle pulled out a base map and measured the distance between the personnel office and the ESD building. He pulled out his slide rule and determined that even following the posted speed limits, his new subordinate should be able to make the drive in fifteen minutes.

At 0915 he called again. "He had to stop at the vault," Staff Sergeant Ellis said. "His security clearances were revoked but they said it would be no problem to get them reinstated."

"Hah!" Boyle said. "Once your clearances get revoked it practically takes an act of Congress to get them back. Don't add him to our roster just yet!" Things were looking up in the world of Major Wesley T. Boyle.

At 0955 there was a knock at the door, a captain in the standard light blue shirt and dark blue pants. "Yes?" Boyle answered.

"Captain Ernest Bellum reporting," the captain said.

"Is that how you are supposed to report to your new commanding officer?"

Captain Bellum looked genuinely put upon but resigned himself to the protocol. He placed his brief case on the ground, stepped forward and stood at attention, and slowly raised his right arm to an approximation of the military hand salute. "Captain Ernest Bellum reporting for duty, sir."

Major Boyle eyed his new subordinate with suspicion. The angle between his right arm and his torso was sixty degrees at best, a full 30 degrees short of regulation. While his uniform was neatly pressed, he didn't bother to wear his ribbons under his wings. Boyle returned the salute. "At ease," he said. "I heard your security clearances were revoked. Why is that?"

"Extended period in a non-allied country," he said. "That's what they tell me."

"How long will it take to get them back?"

"Already done."

"How is that possible?"

"I don't know, but I gave the clearance paperwork to the sergeant out front."

"I see," Boyle said. "Give me a minute. Take a seat."

The captain sat and Boyle walked out to see Staff Sergeant Ellis. She looked up and turned the paperwork around to face her boss's view. "It's all in order, sir. He has the same list of clearances that everyone else in this branch has and a few more." Boyle scanned the list. These security clearances above TOP SECRET had coded names that were themselves classified. The names were usually two generic words that were unlikely to be used in open conversation, such as PATRIOT BLUE or CANON WHITE. There were four such clearances needed for the D-2 branch. Captain Bellum had those and three more that were redacted, since the form itself didn't have the required clearance.

Boyle returned to his office. "It looks like you have the right clearances, captain. Welcome aboard."

"Thank you," he said.

"It is always good to have a real engineer on staff. Even though you're just an MIT grad." The captain didn't react. "What do you know about our branch?"

"I know you work ECM to include detection and mitigation," he said. "I have an interest in both."

"Well, that sums it up. More or less. Right now, we have been tasked with investigating new methods of detection and I am afraid we are well into that, so it wouldn't really help to have someone new jump into the middle of things. How about we assign you to something less intensive until the next project comes up?"

"I don't want to waste my time."

Boyle grimaced. It was bad enough the captain was refusing his suggestion, but he was doing it so rudely. "Listen, captain. Let's get one thing straight. I set the orders around here and while I might ask you for your acquiescence now and then, I am not asking. I am giving you direct orders. And do I really need to remind you that you need to preface, and or conclude each statement to me with the word 'sir?'"

"Got it, sir."

"Fine. Sergeant Ellis will assign you a desk. Ask her for our introduction package, that should take the rest of the day for you. And, by the way, proper attire in this office is the light blue shirt with ribbons. I expect you to be dressed accordingly tomorrow. Bring your Class As. We have an awards ceremony at 1000 sharp. Office hours are 0900 to 1700. Dismissed."

The captain stood, saluted, weakly again, and left. Major Boyle cursed his luck to be handed an insubordinate MIT grad to babysit while he had much more important and pressing duties. It was Boyle's first chance at leadership, and he wanted desperately to succeed. None of his leadership training prepared him for this exact situation.

* * *

The next morning Major Boyle arrived at the ESD building early, dressed neatly in his freshly pressed Class A uniform. The outer jacket was dark blue with his major's insignia perched on each shoulder, his five ribbons over the left breast and a blue nametag on the right. He stopped by his office. The captain's desk was empty.

"Where is Captain Bellum?" he asked Staff Sergeant Ellis.

"He called in and said he has an appointment, but would be at the ceremony on time."

"I have to be backstage before the ceremony," Boyle said. "Make sure everyone is in their seats five minutes early and save a seat for me. I'll need my seat afterwards."

"Yes, sir."

Boyle rechecked that his tie was straight and his five ribbons parallel to the ground as he stood and smiled. Nothing was going to ruin the rest of his day.

The ESD auditorium was a magnificent room with plush seats, an elevated stage, and superb acoustics. Boyle smiled at several early arrivals and made his way onto the stage and then in back, behind the curtains. The other recipients were already in place.

"Wesley, good to see you on time, as always," the lieutenant colonel said.

"Yes, sir, I have been looking forward to it." Boyle shook hands with his fellow award recipients. The lieutenant colonel had an impressive array of five rows of ribbons with three in each row. Fifteen! The other two majors had a paltry four ribbons each, but after today would have five. Boyle smiled at each, knowing he would soon have six.

"Full house," one of the other majors said, peeking around the curtain. "Where's Colonel Lathrop? We can't very well start the show without him."

Boyle checked his watch. 0859. The boss was cutting it close. Major Bennett, Colonel Lathrop's executive officer, rushed onto the stage from the back entrance. "There has been a change," he said. "General Sams is here."

Boyle's face lit up. "General Sams is giving us our awards?"

"No," Major Bennett said. "There will be two award ceremonies. You four are in the second ceremony and need to very quickly use the back door and get to your seats."

"But . . ."

"Right now, please quickly. Colonel Lathrop will call you from the stage when your time comes."

Boyle followed the others out the back, around the hallway, and into the auditorium. After letting his eyes adjust to the darker lighting, he found the D-2 branch seated with an empty seat for him. As soon as he was seated his staff started with the questions, but a voice boomed from the back. "A-ten-hut!"

Everyone came to attention and heard the footfalls of officers walking down the center aisle. Colonel Lathrop was in the lead, followed by General Sams and . . . Boyle turned his head to be sure. Captain Bellum. The three climbed the stairs. General Sams and Captain Bellum sat in chairs facing the audience and Colonel Lathrop stood behind a podium. Beneath Bellum's wings there were four rows of ribbons. Boyle felt his pulse pounding in his temples. It was too much.

"Be seated," Colonel Lathrop said. "Today is a special occasion for us in the Electronic Systems Division. General Sams has flown down from Wright-Patterson Air Force Base to help us welcome our newest member, Captain Ernest Bellum. General Sams."

The general got up from his seat and the audience came to attention. He motioned with both hands for everyone to sit down, "please, be seated." He was six-foot-six inches tall, a full ten inches taller than Colonel Lathrop. The microphone squealed in protest as he elevated it. "I am very pleased to be here today to do the honors for Captain Ernest Bellum's awards ceremony. I met Ernest two months ago, right after he was released from the Hanoi Hilton."

The audience gasped. "Yes, that Hanoi Hilton," the general continued. "Captain Bellum is an Electronic Warfare Officer by trade and has several missions right over Hanoi but on the last mission, a surface-to-air missile blew his bomber to pieces, killing three of his crew of six. And that is why I am here. The reason Captain Bellum is here is that he is also a world class electrical engineer and when asked what he wanted to do for the next chapter of his Air Force career, he said he wanted to fix what was broken in the world of ECM. And I have no doubt he will do that." The general paused and turned towards Bellum and Lathrop. Gentlemen?"

Captain Bellum stood at attention on a masking tape "X" left in the exact center of the stage, with the general on his right. Colonel Lathrop lowered the microphone and spoke. "Attention to orders."

The audience stood at attention. "For exceptional meritorious service while engaged in action against an enemy of the United States, Captain Ernest J. Bellum is awarded the Distinguished Service Medal with a combat 'C' device." Bellum executed a right face turn as the general turned left to pin the medal just below his ribbons.

Boyle's eyes moistened with anger. The captain already had twelve ribbons! And now he was getting another. And it was a DSM! No engineers got that. The officers on stage faced the audience, which started to clap.

"Attention to orders," Lathrop said, interrupting the applause.

Another! Boyle couldn't believe it.

"For gallantry in action against an enemy of the United States during conflict with an opposing force," Lathrop continued, "Captain Ernest J. Bellum is awarded the Silver Star."

"Attention to orders," Lathrop said again. "For wounds sustained as a result of combat against an opposing armed force, Captain Ernest J. Bellum is awarded the Purple Heart."

"Attention to orders." The general added the Vietnam Service Medal and faced the audience and started to clap. Everyone joined in. Boyle added his applause but stared heavily at the four shiny medals hanging from colorful cloth sashes. He knew each medal had an equivalent ribbon. That meant Ernest would show up the next day wearing twelve plus four, sixteen! That meant five rows of three plus one.

As the general spoke, reminding everyone at ESD they existed to keep crewmembers like Captain Bellum safe, Boyle thought anew of his soon to be sixth ribbon. It just wasn't fair. Everyone came to attention and the general left, Captain Bellum alongside, the two chatting easily.

"Be seated," Colonel Lathrop said. "That isn't the only excitement we have for today. Will today's award recipients please come forward."

The rest of the day was spent on what aviators called autopilot. Major Boyle ended the day accepting the congratulations of his staff and wondered about his policy that all officers must wear their ribbons on their light blue shirts. He couldn't make up his mind. The next morning, Captain Bellum sat at his desk without his ribbons. Major Boyle closed his office door, removed his ribbons, and massaged the holes in the fabric where the posts to the ribbons used to be. He opened his office door and called out, "Captain Bellum, front and center."

Bellum appeared at the office door. "Yes, sir."

"What do you want to do that won't waste your time?"

"I want to come up with an automatic jammer."

"Automatic?"

"Yes, sir. You turn the thing on, it determines the frequencies, jams the threat and switches frequencies as quickly as the bad guys can."

Boyle thought for a moment, realizing it was perfect. "Okay," he said. "You do that. Don't collaborate with anyone in this branch, they are too busy. You are solo. Also, don't bother me with it until you are done. Dismissed."

The captain returned to his desk. The next morning, nobody in the office wore their ribbons.

46. April 4, 1974

Concord, Massachusetts

"Officers, center face!" a voice bellowed as the wedding guests looked on. Eight Air Force officers, the ushers, four on either side of the pathway leading from the church, executed right and left face turns to face inward. Each stood at attention.

"Officers, draw swords!" the voice pealed again. Each of the eight ushers unsheathed their swords and pointed them skyward, and then lowered them, angled inward with their cutting ends up, forming an arch.

"Ladies and gentlemen," the baritone sung out again. "May I present to you, Lieutenant Marjorie Bellum and Captain Ernest Bellum." From the church door, Marjorie and Ernest walked through the saber arch. He was dressed in his full mess dress uniform, she in a white wedding gown. They walked slowly, as rehearsed, in what most in the military would recognize as "half step march."

As they neared the last row, the last two ushers lowered their swords, blocking their path. "Give him a kiss!" one of the ushers demanded.

Marjorie faced her groom, perched on her toes, and kissed Ernest as the cameras snapped away and the crowd cheered. The last two ushers raised their swords.

As protocol demanded, the couple led the way to the reception venue. It was a quick three-minute walk from the church to a large outdoor tent, erected outside the historic Colonial Inn. It was a brisk spring day, but the skies were clear, and the winds were light. The morning weather report showed extensive tornado activity in the Midwest, but none of that impacted

Concord, Massachusetts. Stephen, as the Best Man, stood at lead of the reception line, followed by Ernest, Marjorie, and Caroline, Marjorie's sister and maid of honor. Each guest handed Stephen a card, and he introduced the newlyweds.

"Mister and Mrs. Ralston, may I present to you Captain Ernest Bellum and your daughter, Lieutenant Marjorie Bellum."

"Ernest that was spectacular," Katherine Ralston said. "I've never been to a military wedding before. And you look so handsome in your uniform." She hugged Marjorie. "And you, my dear. You are so beautiful. I am so happy you decided on the gown."

"Me too, mom. Ernest insisted."

"Smart man," Roger Ralston said, shaking Ernest's hand so tightly the pain radiated up his right arm, causing him to grimace. "Oh sorry."

"Not a problem," Ernest said. "The docs say I'll grow used to it."

"Easy for them to say, isn't it? Can't they do anything for you?"

"Not really. They say having it untreated for as long as it was, things are just finding a new equilibrium. I'll learn to live with it."

"We are very grateful for the U.S. military, Ernest. Our country is what it is because of men like you. And women like Marjorie. We are very proud."

"Thank you, sir," Ernest said. "You have a wonderful daughter."

"And you now have a wonderful wife."

With each subsequent guest, Stephen warned the males about Ernest's injured arm and the reception line moved smoothly. Once everyone was through, Stephen asked the couple to linger outside for a few minutes longer while he made final preparations inside.

"I know this isn't your cup of tea, Ernest, but I am really enjoying it. Thanks for letting us do this up big time."

"My mom would never let me hear the end of it if we didn't."

"She is a lovely woman, I am so happy with all of your, I mean our family."

"Well, most of us, anyway."

"So, Martin is definitely a no show then?"

"Afraid so. He says he's too busy. He hasn't seen any of us since his graduation last year."

Stephen stepped out of the tent. "It's show time, you two."

They watched as Stephen approached his chair at the head table. He tapped his wine glass with a knife from his place setting. The crowd rushed to their seats and fell silent. Another uniformed guest spoke with a deep baritone voice, "All rise for the arrival of the bride and groom, Lieutenant Marjorie Bellum and Captain Ernest Bellum."

The crowd came to their feet and clapped, politely at first and then with a volume that grew as the couple entered. Ernest blushed while Marjorie beamed. Once they took their seats, Stephen said, "Please, everyone, be seated."

"Welcome friends and family," Stephen said. "We have a feast for you this afternoon, the finest New England has to offer on this, hearty New England day. Please enjoy and get to know your neighbors. We will all get to know more about our wedding couple after dinner."

With that, Stephen sat, and waiters appeared from the margins of the tent to serve everyone, starting with the head table. "I'm too nervous to eat," Marjorie said to her groom, "you'll have to help me."

"That's why I am here," Ernest said. "I will always be here for you."

Susan, seated to her right, diverted Marjorie's attention as she was about to speak. Ernest looked to his left, where his brother was talking to Caroline. He felt, for a moment, as the most awkward groom in history. Stephen took notice.

"How's life at the lab?" he asked.

"I think I am making progress. I have an idea to automate what EWs do to make it faster, more reliable."

"You are putting EW's out of business?"

"Why not? Even at my best, I can only detect, interpret, and jam a new signal every two or three seconds. A computer can do that faster and without relying on the skill of the operator."

"How much?"

"Less than a million a copy."

"Well, I will look forward to that, not that I will be flying any day soon."

"Is staff work agreeing with you?"

"No, but it is a necessary part of career development. We pilots got to pay our dues if we ever want to take command."

"Glad I'm not a pilot. So, what have you been doing as a staff weenie?"

"My job has been to collect all the Linebacker Two after action reports and put them together for an official historic record."

"Still?"

"Still."

"So, you get to write history?"

"I wish that were the case. I get to run an accuracy check on what others have written. And it isn't good."

"How so?"

"Everybody, except the 17th Air Division Commander, are saying it was a resounding success, the first three days were bloody, but the plan worked."

"Sounds like absolute B.S."

"It is absolute B.S."

"What does General Silvers say?"

"Nothing at all. I think he's been told to shut up and color."

"I heard he was the only general officer with any credit in all this."

"You heard right. After Night Four he informed the SAC Operations Staff that his wings would no longer fly the old plan. He didn't ask permission. He informed. The higher echelons had no choice but to give him his way. He saved the day."

"I bet they got even with him."

"Too soon to tell, but I bet he never sees another promotion. He really deserves credit for turning the ship around."

"But history will never know that. It will be recorded as a win for SAC."

"Not if I can help it."

"Enough shop talk, you two!" their mother interrupted, standing between the two. "Ernest that wedding was lovely. You look so handsome dressed like this. Like your father when he was your age."

"I can't take any more gushing over my little brother," Stephen said. "Mom, you ought to get back to your table. I have a speech to give."

Stephen stood and hit his wine glass with a knife twice. The baritone voice from the guests below boomed anew. "Guests and members of the wedding party, provide silence for the best man." After everyone was seated, Stephen spoke.

"When my brother Ernest asked me to be the best man at his wedding I was confused. I said, 'me?' He said, 'yes, you are my second-best friend, and my best friend is disqualified for the task due to her gender.' I told him I was surprised he had any friends at all."

The audience groaned in protest. "No, I am not serious. It is obvious from all in attendance that Ernest and Marjorie are loved and admired by many. I would like to pay special attention to two friends in particular. They were crewmates of Ernest during the first three nights of Linebacker Two. Ernest and these two are rare members of the club that have one less landing than takeoffs. And they shared time together in that institution we know today as the Hanoi Hilton. Major Walter Stevens and Captain Harry Broadmore, please stand and be recognized."

The two stood. After the applause, Major Stevens spoke. "Point of order to the Best Man."

"The point of order is recognized."

"In honor of our lost crewmates," Stevens said, "I command that all in presence charge their glasses." Those well versed at these affairs helped the others to ensure their glasses were full. Once that was done, Stevens continued. "That awful night over Hanoi, whereby through the Grace of God, Ernest, Harry, and I were delivered from certain demise, our three crewmates paid the ultimate sacrifice. A toast." The entire assemblage rose to their feet. "To Captain Kevin Long." He paused. "To Captain Tony Smallwood." Another pause. "To Staff Sergeant John Polanshek. You have not died in vain. Your sacrifice has permitted others to live. We shall never forget." Stevens lifted his glass.

"Hear, hear!" Everyone toasted.

"Point of order to the Best Man," Captain Broadmore said.

"The point of order is recognized."

"To the United States of America," he said.

"Hear, hear!"

"Point of order to the Best Man," Ernest said.

"The point of order is recognized."

"To fallen comrades," he said.

"To fallen comrades" the crowd echoed, somberly.

"Please be seated," Stephen said.

Stephen continued. "In case you were wondering why we are having a wedding on this blustery spring day, on a Thursday no less, let me tell you this is no ordinary Thursday. It was one year ago today that Ernest was reunited with Marjorie and the rest of our family after his months as a prisoner of war. It will be a day of the year we will forever celebrate."

"It is the best man's duty to say a few embarrassing things about the groom and to welcome the bride to her new family. I am afraid I will fail at the first. To be honest, I grew up thinking my little brother was a little weird. While I was doing normal boy things, like playing baseball, he was taking dad's car apart." Jonathan laughed from his table saying, "It's true, it's true."

"When I was twelve, Ernest was eight, and we lived in Germany. As kids we picked up the language well enough to count to ten, play with our friends, and to get along on base. Ernest went missing one day and it was my duty to find him. After looking at all the obvious little boy places I looked at the not so obvious places. I found him at a local radio store talking to a German national, in German, about how to build a radio. In the next few weeks, Ernest built an A.M. radio. He was eight."

"Seven," Ernest said.

"He was young, let's put it that way. He's always understood electronics better than anyone and when he got to the B-52, I thought it was a perfect match. As Major Stevens and Captain Broadmore will tell you, he is very good at what he does. I used to think Ernest was a little weird. Now I understand he isn't weird. He is just so smart, the rest of us can't begin to comprehend what he knows. He is, very, very smart. And I have further evidence of that. She is seated to his right."

All eyes turned to Marjorie.

"Marjorie, it sounds like it was love at first sight. Well, let me tell you, that for our family, your new family, it was love at first sight. Welcome."

"Speaking of family, many of you know our father, Colonel Jonathan Bellum. Dad is known for his rules of life, things he drilled into us boys to the point we had them memorized. I'd like to quote one for you in closing. It isn't the one you are thinking of. So please charge your glasses one more time." Everyone did so.

"A toast." Everyone rose. "The gift of life must be appreciated to be lived to its fullest. Here is to life."

"Hear, hear!"

47. April 4, 1975

Bellevue, Nebraska

It was the usual Bellum gathering on a cold spring day, but this time with the addition of Ernest and Marjorie visiting from Massachusetts. After dinner, as usual, the women lingered over coffee at the dining table while the men sat at Jonathan's fireplace. Once the senior Bellum had finished building the traditional fireplace pyramid – four large logs in a square at the bottom followed by smaller logs and squares – he topped it off with a large log on top. It would be the bomb to drop after the lower levels had collapsed. Then, also traditionally, the men sat and watched quietly until one of them finally spoke.

"Congratulations, Ernest," Stephen finally said. "Promoted two years early, I imagine it will take a while to pin on."

"Two years, at least," Ernest said. "It's for the best. My commitment ends at the end of this year and that gives me some time."

"Getting out? Really?" Stephen asked.

"I don't know, but it's good to have options."

"But why are you even considering it?" Jonathan asked. "As a major you know you can make it to twenty years and have a pension. I know it isn't much, but with the economy the way it is?"

"I feel like I'm wasting my time, dad. I don't know how much I can say."

"I still have my clearances," Jonathan said. Ernest looked to Stephen.

"He does," Stephen said. "I checked."

"Okay," Ernest said. He considered his words while considering the fire and the crackling sounds of the logs, now fully engulfed. Ernest's clearances still exceeded that of his brother and father. But he needed to talk. "I've spent the last two years working on an automatic jammer. I finally got clearance to talk to the vendors of the old ALT-6B and ALT-22. Come to find out, they already had an automatic jammer ready to go before Linebacker II. SAC refused to fund it, saying the money was better spent on facilities improvements."

"How much?"

"A million a copy," Ernest said. "We lost fifteen B-52s and how many KIA?"

"Thirty-three," Stephen said. "Plus another thirty-three captured and I think twenty-six MIA."

"I wonder how many could have been saved?" Ernest said. "And they let me waste my time designing something they already had. My boss knew about it but let me reinvent the wheel just to keep me busy. The dumb thing is that even if you didn't care about the lives lost, the cost of the bombers would have paid for the program ten times over."

"It could be that the people running the show today didn't know about the technology invented a few years ago," Stephen said.

"Could they be that stupid?" Ernest asked.

"You bet they could," Jonathan said. "But it is more likely that they don't want to acknowledge that the new jammer existed back then because it would expose their bad decision not to fund it. Happens all the time. Tell him, Stephen. Tell him about the after-action report."

"Yeah, that is true," Stephen said. "Eighth Air Force put together their after-action report saying SAC's Linebacker Two plan was practically flawless. They predicted a three percent loss rate. If you take the fifteen aircraft lost and divide that by the total number of missions, it comes to just over one percent."

"I thought it was higher than that," Ernest said. "It seemed higher."

"It was," Stephen said. "The losses after the first three nights came to over six percent. Then, after General Silvers refused to continue the old plan and SAC gave up control, Eighth Air Force finally had the guts to change the plan. It was only then, starting on Night Five, that the loss rate fell."

"Can you get this into writing?" Ernest asked.

"I wrote a dissenting report," Stephen said while rising and then kneeling at the fire. He used the prong of a poker hanging on the mantel to move a log that had fallen into the center, knowing his father would want it that way. Jonathan had removed his prosthetic leg, revealing a rash around the stump around where his knee used to be.

"What chance does a dissenting report from a major halfway around the world have against that of a three-star who was in theater?"

"Not much," Stephen said, retaking his seat. "But you got to try, don't you?"

"Yeah. But if that fails, the winners do write the history," Ernest said. He watched as the freed-up space in the log pyramid opened the air flow and the glowing embers turned into flames anew. "Speaking of history, has anyone heard anything at all from Martin?"

"Not a peep," Jonathan said. "We got a newsletter from the college that said he's now a professor of history. Just like we knew he would be. We send him letters; he never writes back. We leave messages at his office and his secretary promises he'll call back. He never does."

"Time for a hospitality check then," Ernest said. Stephen laughed.

"What's that?" Jonathan asked.

"I guess that was after your time as a crew dog," Stephen said. "These days, if a crewmember is getting unsociable, it is up to the rest of the squadron to show up unannounced, en masse, and throw a party at the hermit's home. Force him to be a part of the squadron."

"I can't imagine that going over well with Martin," Ernest said.

There was laughter in the next room. "It sounds like not everything is doom and gloom in the family," Jonathan said. "Thank God for the Bellum women."

* * *

The Bellum women were each seated at the dining room table, each with a cup of coffee. Susan was describing her latest conference with her eldest son's teacher. "'What does your dad do?' the teacher asked. 'He's in the Air Force,' Justin had answered. 'Does he fly airplanes?' 'No, he pushes paper.'"

"Stephen was unhappy to hear that," Susan confided. "He's thinking about joining the local aeroclub just so his children know that he is an actual pilot. Trouble is that he doesn't have the license to fly something with only one engine."

"It must be rough," Marjorie said. "He is a pilot and an Air Force officer. It sounds like his self-identity hinges on both aspects of his career. Or should I say careers?"

"It is difficult. He has five more years before he can retire with a pension. But he is looking at the airlines. Maybe."

"No!" Mary said. "Give up his pension? I won't hear of it."

"He hasn't decided yet, mom. But the economics work out. After a few years with a major airline, the amount of the pension is just a fraction of what he would be making. But he's still not decided." As she spoke, she rubbed her finger along a chip in the formal coffee cup. Susan had been a frequent guest at the senior Bellum's home, well before Jonathan had retired. She remembered when the china set was brand new. Stephen and she talked often about the paltry sum of an Air Force colonel's pension.

"Well, I guess if you have thought it through," Mary said. "It seems like the Air Force is part of the family after all these years. But it really isn't."

"How is Ernest doing?" Susan asked. "We hear about the terrible psychological toll on POWs. But he seems perfectly okay. It couldn't have been easy."

"I don't know if he's perfectly okay," Marjorie said, emphasizing the word 'perfectly.' "But he seems just like he was before the Hanoi Hilton. He still can't lift his arms above his shoulders because of the way they had him tied up for the first few days. The base doctors here say he spent so many months without getting the shoulder joints treated, the bones have fused into new sockets. But other than that, he doesn't seem to have any lasting physical issues."

"What about mental?"

"After we got married, Ernest was 'textbook normal,'" Marjorie said. "I know that sounds weird for someone I know people think is textbook abnormal on a good day. But he really has a wonderful sense of humor about everything. You've heard him talk about his buddy Nicholas. 'Neekolas the Rooskie,' he calls him. I used to think he managed to get through the entire POW experience without any emotional baggage at all."

"You used to think that?" Mary asked. "But not anymore?"

"Things are not so good at work right now," Marjorie said. "It's all top secret so he can't talk about it with me. But I can tell. It is eating away at him. He can go days without a smile."

"No," Mary said. "That isn't right."

"He's fine with me, he tells me so," Marjorie said. "I believe him. When I look at him with my 'I'm worried' expression, he just says he's 'having a moment.' And then things go back to normal again. But he is having more and more of these moments lately. I just don't know what to do. I was thinking about seeing the base psychologist. But one of the wives from his office told me if I did that, he would lose his security clearance."

"What can we do?" Mary asked.

"Oh, just be wonderful, as you always are. I love him dearly and I know he loves me. I am thankful we are assigned to the same base. I think it will all work out. I know it will."

"Everything happens for a reason, dear," Mary said.

"That's what Ernest says," Marjorie said, nodding. "I was starting to believe that."

48. What is War Good For?

Berkeley, California

Martin sat at his desk, considering the two stacks of mid-semester term papers facing him. He had just finished reading through the first stack and the work of his sophomore students was not up to his expectations. The newly arrived stack, from his senior level students, awaited. His secretary knocked at his open door. "Professor Bellum, you have just enough time to make it to the lecture hall."

"Thanks, Stephanie, I was lost in my thoughts."

"Again," she said, smiling. She wasn't his secretary *per se* since he shared her with other non-tenured professors, but he liked to think of her that way. She was single and attractive. He was single and lonely. But he hadn't yet mustered the courage to ask her out.

He grabbed his notebook and headed for the door, just as Eli Saunders, his officemate, was also leaving. "Heading to building three?" Eli asked.

"Yeah, I can just make it."

"I'm headed to number two. Life as a first-year college professor isn't quite as glamorous as I thought it would be."

"I don't know," Martin said. "It beats living in a dorm and sweating out the draft."

"There is that. Where you living?"

"I got a one-bedroom apartment just off campus. I can walk to work."

"Nice! How many in your lecture?"

"These are the sophomores, so, four hundred, give or take."

"Nervous?"

"Not anymore. It made me nauseous the first few times, especially this large an audience. But I actually enjoy it now."

"Maybe I'll get there soon. You have two a week, as I recall."

"Yeah, Wednesday and Thursday. I have four graduate students doing the classrooms and they are great."

"You still reading every paper?"

"Yeah, I actually like that bit. Good kids."

"They do seem to be more serious these days."

"They are. Kids these days are more focused on their educations. I know we still have people in Vietnam but getting the bulk of the Army out has done wonders to heal the country."

The neatly manicured sidewalk forked into two, the left leading to the humanities building, the right to the history building.

"Here's me," Eli said. "See you later."

Eli peeled off toward Building Two of the campus Liberal Arts quadrangle, leaving Martin to continue alone. He checked his watch. He had just enough time to duck into an adjacent faculty lounge, catch his breath, and then saunter into the lecture hall with the nonchalance that telegraphed to his students that he had been doing this for years.

As he walked into the hall the students dutifully became quiet and directed their eyes to him. The lecture hall seats started at his level, near a large chalkboard, and then radiated upward, ten levels. There were a few empty seats, but not many. He was happy to see the teaching assistant on duty had written "HIST 203 The True Cost of War" and "Prof Bellum" on the chalkboard. He placed his notebook on the podium, stepped to one side and looked at his young students.

"War!" Martin bellowed. "What is it good for?"

"Absolutely nothing!" most of the class yelled back, as if on cue. It was going to be a good class, Martin knew. He spoke mostly from an outline of points he wanted to make, only turning the page when making a specific point from one of the assigned readings. He ended as he had begun. "War! What is it good for?"

"Ending fascism!" a planted student yelled out. The audience appeared stunned.

"Was it really war that ended fascism?" Martin asked with a smile. "As you dive into the next set of readings, I want you to think about that and consider two more questions. Could it be that fascism ended on its own accord? Or could it be that fascism didn't end at all? These are the questions we historians need to consider."

Martin turned to the chalkboard and wrote in large letters: FASCISM, IMPERIALISM, COMMUNISIM, MARXISM, COLONIALISM.

"Your term paper for this semester will answer one of three questions. First, you can have the premise that whatever the outcome, that the -ism in question could have been solved in another way. Second, you can also argue that the -ism of your choice isn't something to be afraid of but should have been embraced. In that case, you need to make the argument that war is either good or bad for achieving the -ism. And finally, you have a third choice. You can have the premise that war is good for ending one or more of these -isms. If that is the angle you choose, be forewarned that history is against you. Your TAs will have more for you. See you next week."

Martin collected his notebook and walked out, just as the hallway bell rang out, denoting the allotted fifty minutes was over. The students would have to scurry along to their next classes while Martin had the rest of the day off. *A master of my craft,* Martin thought as he left the building and the warm California sun hit his face.

Strolling along the sidewalk he spotted his old hangout, where he and Sherrie spent hours drinking coffee and considering the ills of the world. "A celebratory coffee?" Martin asked himself aloud. "Why not?"

It was nearing lunchtime and most of the booths and tables were taken. He found a spot at the bar where Derrick, the café owner, greeted him. "The usual, professor?"

"Yes, please, what's that?" Martin asked, pointing at the television on the wall.

"Saigon," Derrick said. "They say this was yesterday but with the time difference we're only seeing it now."

A helicopter was perched atop a building while a long line of people fought their way up from the ground level to ladders perched precariously along the walls and the building itself. The building rooftop was barely big enough for the landing skids of the chopper and the line of people on the ladder appeared to be more than the helicopter could safely carry. There were swarms of helicopters in the distance and the sounds of bombs closer by. As the helicopter took off, one of the stragglers grabbed onto the skids, only to fall four stories below.

"He really wanted on that chopper!" Martin said.

"This is the end," Derrick said. "Anyone they don't evacuate ain't going nowhere anytime soon. You hear about the plane crash?"

"No."

"An Air Force C-5 crashed trying to evacuate everyone they could carry. Killed about a hundred and fifty, half of those were children."

The news story shifted to the scene of an American aircraft carrier packed with people on the deck. After each helicopter landed and its occupants disembarked, the chopper was pushed overboard into the sea. "They are making room for as many as they can physically fit . . ." the news reporter said.

"I don't understand," Martin said. "Those are Vietnamese fleeing their own country. It doesn't make any sense."

"I have one of your students bussing tables here at night. Ha Tàu is her name. She still has family there. You should talk to her."

"I know that name," Martin said. "I've read her papers. She's Vietnamese?"

"On a student visa. She graduates in May and was supposed to go home. I wonder what she'll do now."

49. Not How Things Are Done

Lexington, Massachusetts

The best thing Ernest could say about his boss when he started at the Electronic Systems Division was that Major Boyle was "hands off." He didn't want to know what Ernest was doing and gave him no direction at all. When it came time for his first annual Officer Effectiveness Report, Boyle had given Ernest a less than perfect rating. Less than perfect in the current officer climate was the same as saying, "do not promote this officer." Rather than sign the OER, Ernest sent it to Colonel Lathrop, the division chief, with a note saying, "what should I do about this?" The next week the OER was written to the perfect level. When Ernest was promoted two years early, Major Boyle graduated from being unfriendly to openly hostile. In a few years they would both be majors; Boyle would need to come up with another plan.

"Write it yourself," Boyle said when it came time for the next OER. Ernest did so, saying he was working on new ordinance delivery methods which made electronic counter measures unnecessary.

"What happened to your automatic jammer?" Boyle asked.

"You knew all along," Ernest said. "The manufacturer says you were briefed when you took over here."

"Yeah, I guess I forgot that. But tell me about this new program."

"It's turning high altitude ALCMs into low-level cruise missiles," Ernest answered.

"Can never work, we already looked at this. You can't pack enough fuel into them when flying low and the inertial navigation systems just aren't good enough. You should know that." That was certainly true of the first generation of Air Launched Cruise Missiles and the progress in the next generation was slow in coming. Pronounced "al cam," the program was a hot project a decade prior, but appeared to have lost steam.

"The current ALCM was designed in 1960. We can double the range even at low altitudes with new technology engines. I think we can computerize terrain maps and have the missile navigate using that."

"No inertials?"

"We probably need to keep them, but I am looking at alternatives."

"How much?"

"A million a copy."

Major Boyle whistled. "You want to spend a million dollars per missile?"

"It cheaper than losing a bomber."

"Maybe it would be cheaper to improve the ECM."

"Why not do both?" Ernest watched his boss's eyes as he thought. They always tended to dart to-and-fro and rarely kept "lock" onto whomever he spoke to. But when he started to fabricate the next lie, they settled straight ahead and focused on the lie's recipient. He looked Ernest in the eyes.

"This isn't how things are done around here. The ideas start from the top and filter their way down."

"Then things are done incorrectly around here."

"We'll see about that. You will hold all work until I get this approved or disapproved."

Ernest returned to his desk and waited for the next victim to distract Major Boyle. As soon as his boss's door closed, he picked up the phone. "Bob, it's Ernest. I'm ready. How about this afternoon, say at two? Perfect, see you then."

Ernest gathered his notes and cleaned his desktop. Major Boyle normally left for lunch at 1145 except on Tuesdays, when he dined with his boss, Colonel Lathrop. On those days, Boyle departed at 1130. Boyle signed out on the chalkboard above

Sergeant Ellis' desk at 1130. Ernest signed out at 1145, "MIT – ROD," meaning he would be gone for the rest of the day.

* * *

The Lincoln Laboratories at the Massachusetts Institute of Technology were hidden away in a forest right next to Hanscom Air Force Base but required a short drive out of the main gate and down to the aptly named Wood Street. Security was as tight as the Air Force Base and in some buildings even tighter. Ernest presented his credentials to the guard who picked up a phone. Within a few minutes, Bob Jenkins appeared and started the process of getting Ernest signed into the building.

Bob was barely five-foot-five and couldn't have been much heavier than 130 pounds even with a heavy winter jacket. Because they were usually seen together, their college classmates called them "Mutt and Jeff," with Ernest playing the role of Mutt. It was widely assumed they were best friends, but they rarely spoke during their four years together. Common interests meant they took the same classes, but both tended to be loners. It wasn't until recently that a friendship could be said to have started.

Bob wore a wrinkled white shirt, black tie, and wrinkled blue slacks. He handed over the last clipboard of paperwork to the guard and Ernest was waved in. "Howdy," Bob said in his exaggerated Texas drawl. "Let's get into the vault."

Ernest followed Bob through the labyrinth of the Lincoln Labs to one marked "Center for Systems and Architectures." Bob negotiated with a mechanical door lock which agreed to their entrance with a metallic buzz. The laboratory was a maze of its own, but they ended up at Bob's desk. "Now then," he said, "can we finally get on with this?"

"Not just yet," Ernest said. "My boss finally asked what it is I have been doing this last year so, as we agreed, I told him. Now he says he is going up the chain of command with this."

"Will he?"

"I don't know. I know his motives are to promote himself, but this doesn't do that. He also wants to avoid having his subordinate cast in a good light without sharing some of the glow. There is a chance he'll approach the colonel."

"Lathrop?"

"Yeah, Lathrop. That's where it gets interesting. Colonel Lathrop wants to be General Lathrop. This project isn't in his budget, which means he has to ask for more."

"He's not going to do that, is he?"

"Not a chance," Ernest said, smiling.

"You don't seem too broken up about that, Ernest. Does that mean what I think it means?"

"Yes."

"When can you start?"

"I need to talk this over with Marjorie first. But I think she'll be all for it. I have to sign a form accepting my promotion no later than the end of this month, or I have to formally resign my commission."

"And?"

"And, once I do that, I will be a civilian at the end of the next calendar month."

"That's great, Ernest. Let's get your paperwork on this end started."

50. Temptation

Papillion, Nebraska

Stephen pulled up to his house just before the taxi and in time to meet his guest. "Great to see you, Jack!"

"You too, Stephen. I got to tell you; you look ten years older. SAC hasn't been kind to you."

"That's an understatement. I feel ten years older. Let's get in and you can meet Susan."

Stephen led the way down a short hallway and down four stairs to the bottom level of their split-level home. Susan was in the kitchen but came out to meet the men halfway. "Susan, this is Jack Haverill, Jack this is Susan."

"I've heard so much about you," Susan said. "I guess you got out of Saigon in the nick of time."

"Nothing so dramatic, I think Stephen might have embellished my exit. I actually rotated out when my tour was up. It was after the peace treaty but at least a year before the city fell."

"Well, it is nice to have that nightmare over for all of us," Susan said. "The kids will be home from school soon. Jack, can I offer you anything to drink before I get busy with the homework and dinner routine?"

"No thanks," Jack said, following Stephen's lead to the living room.

"How is life as an airline pilot?" Stephen asked once they were seated.

"Good, good. Well, fantastic actually."

"No regrets punching out early?"

"No, the Air Force actually made it easy. They promised me back to a cockpit after I did my two years doing the staff thing and rewarded me with an instructor pilot position in Enid, Oklahoma."

"Ouch."

"It all worked out. I interviewed with United Airlines the week after I came back and got hired before the postwar rush."

"How's the pay?"

"I just made parity, so it hurt for a year. But I will be making double this time next year. Plus, I get a staff override bonus."

"What's that?"

"Most airline pilots don't want to do anything but fly, so they have to entice pilots to do the staff work that requires a pilot. So, I work in the personnel department screening pilot resumes five days a week except for the days I pick up a trip."

"Don't you get paid by the hour, though?"

"That's the neat thing about it. As a staff guy, I get paid for 75 hours of flying each month, even if I only do a fraction of that. We moved to Denver, where the main training center is. I'm so junior, I can only hold a line as a Boeing 727 flight engineer in Denver."

"Sitting sideways."

"Yeah, so I'm not actually in a pilot's seat for now. So, I'll do the staff thing until I can bid first officer. I work less, get paid more."

"I think we got that backwards in the Air Force, let me show you something, I'll be right back." Stephen left briefly and returned with a stapled report. "Here, read this. I'll get us some beers."

Stephen returned with two beers as Jack turned the second page. Jack accepted the beer and tilted it in salute. "Cheers."

"Cheers."

"Is this official?" Jack asked, raising the report.

"Not yet, I'm fighting it."

"What are your odds?"

"Not good."

"It makes it sound like SAC planned and executed Linebacker Two brilliantly. Well done, all around."

"I know."

"There are enough of us out there that know what really happened."

"The winners write history, Jack, you know that."

"You think Vietnam was a win?"

"Like I said, I'm fighting it."

"And if you lose?"

"I don't know."

"Give me a call. I can't get you hired, but I can get you an interview."

"Me? An airline pilot? What are the odds they will hire a bomber pilot who hasn't touched an airplane in three years?"

"Better than you think."

51. Ha Tàu

Berkeley, California

Martin had to read the paper twice to be sure. He had given his students three options, but the last option was just a joke. What he wanted were term papers that verified that all wars are wrong, and his preferred answer was that the -isms the war was fought for could have been defeated in another way, other than war. He would also have gladly accepted arguments that the -isms themselves were not wrong, just misunderstood. But what he didn't expect were any papers choosing the third option: an example where war was the only way to end an -ism. And yet there it was.

He knew of the student because she was the only Asian in his senior lecture and had a name he couldn't forget: Ha Tàu. After his first reading of her paper, he asked his secretary to print out her records. He was surprised by her age, 25, a bit old for a college senior. But she still looked younger due to her diminutive size, long black hair, and clear complexion. The words on her paper did not fit the vision he had of her face.

Ha argued that the Vietnam War could have ended with a free South Vietnam, allowing greater freedoms for her countrymen and providing an example for the North about how a democracy was better than a communist autocracy. She argued that Linebacker II had broken North Vietnam's will to fight further, but that the Americans had failed to follow through. In her view, the war was not only worth the fight and the right thing to do, but it hadn't lasted long enough. Given another year, South Vietnam could have defended itself on its own and the North Vietnamese would have no other options.

Martin was unprepared to challenge any of her assumptions, but he knew he had to try. Ha had somehow become corrupted by her time in the U.S. He made a copy of her paper and returned the original to the stack with a note clipped to it. "Schedule a time so we can talk about this."

When Ha appeared on his schedule, he promised himself he would research some of her assertions, but whenever he started, he found he couldn't continue. The subject was just too painful in his personal experience. The day he walked into his office and saw her waiting in the waiting room outside his office, he felt totally unprepared. But there she was.

"Hi, come on in."

She was dressed in jeans and a red blouse that drew attention to her olive complexion and jet-black hair. He thought she was exotic, but then chastised himself for categorizing her into a stereotype he felt was unfair. All these thoughts, he realized, made him feel uneasy in what he had to do next. Ha was in the wrong; he knew in his heart that he was in the right. It was she who needed to be saved, not he.

"You wrote a very good paper," he began, "but I'm not sure I agree with your conclusions." She listened quietly, her eyes locked on his, but her expression was of kindness, not defensive at all. He thumbed through his copy of the paper, which he had mercilessly written over in red ink. "For example, you write that the people of South Vietnam are worse off now than they were when their country was occupied by the Americans. How long has it been since you have been away?"

"Four years now," she said.

"So how can you know they are worse off now if you haven't actually been there since?"

"I have family still there," she said. "A father, a mother, and two sisters. Only my brother and me are here now."

"And do you communicate with them?"

"Yes, but not with my parents since June."

"Why is that?"

"Because they have been imprisoned," she said, her eyes moistening. Martin wasn't expecting this, and his margin notes suddenly seemed inappropriate.

"Why is . . ." he started to ask.

"My father was a town clerk in Saigon, so he was classified as an official of the former government. He was just a typist, that's all. But all members of the government have been sent to Tra cài tao. That is what they call re-education camps. They are not permitted visitors. My sisters say they have seen him in work groups, building roads and doing manual labor."

"And your mother?"

"Nobody knows."

"How is life for your sisters?"

"They are younger than me and are both unmarried. The government seized our house and moved them to work camps. But we are not allowed to communicate."

"So how do you know all this?"

"Our neighbor escaped on a fishing boat. His family gave him all the money they had so he was able to get to the Philippines and there he contacted the U.S. embassy. He has sent me many letters. It is so, so heartbreaking." Now she was crying, but she hid her eyes with her hands. Martin grabbed a box of tissues and handed it to her. She wiped her face and looked up, still silent.

"I'm sorry," Martin said. "I don't know if I can critique your paper under these circumstances."

"No, you must. I wish to pass your course so I can graduate."

"Please don't worry about that. What will you do after you graduate?"

"I don't know," she said. "My Visa expires, and I am told I must go home. I don't know."

"The U.S. is accepting refuges from Vietnam, aren't they? There must be a way you can stay."

"I hope so. But I hope also that I can bring my parents here."

"Have you talked to anyone?"

"The campus foreign student office tells me I must go to the State Department in Washington, D.C. But I am told I cannot be seen until my Visa has expired, not before."

"Well, that's not right!" Martin said, realizing he raised his voice. "Let me make a few phone calls. We have class on Thursday, right?"

"Yes."

"See me after the lecture, just come to the front afterwards."

"Thank you," she said, smiling for the first time. "And what of my paper?"

"Your paper is great," he said. "A-plus."

"Thank you so much. I shall see you on Thursday."

Martin spent the rest of the day on the phone with several offices on campus and the answer had become clear. "You need to talk to the State Department."

After another day on the phone, Martin had what he thought might work. When Ha approached him after the lecture he didn't know where to begin, so he just started talking.

"I booked us tickets to Washington, D.C. next week. It is spring break. You have the week off. Right? Because this is your best shot. We'll leave on Monday, be back on Wednesday."

"No. What? I cannot go. Who will pay?"

"I will pay," Martin said. "I have a classmate from undergrad who works in the State Department. He says the Deputy Chief of Foreign Affairs for Southeast Asia is sympathetic to causes like this and he put us on his schedule. We have to do this."

"We? Why we? Why are you doing this for me?"

"Because," Martin started to say and then stopped, realizing he didn't have a reason. "I don't know," he said finally. "It seems an injustice was done to you and your family. I feel like I should do something."

"Will it work?" she asked.

"I don't know. But what else can we do?"

"Okay," she said, finally.

52. Proof of Concept

Lexington, Massachusetts

"After all he's been through? He deserves better."

"He really does, mom," Marjorie said. She stood in the kitchen, holding the phone in one hand and a reheated casserole in the other. The clock on the oven read 9 p.m.

"He says he'll be home soon. This is pretty typical lately, I'm afraid."

"What is so important to keep him at work all day?"

"He's working on something that should prevent bombers from ever getting shot down again."

"How?"

"I don't know, he's not allowed to talk about it. We don't talk much about anything these days."

"Oh no!"

"No, it isn't like that. Ernest gets fixated on things and you can tell he's thinking all the time. That brain of his is never idle. He just needs some time."

"Does he have anytime left over for you?"

"Sure, a little." Marjorie fell silent, searching the archives of her brain for any evidence to support her claim. "We walk the dog around the neighborhood together. On the weekends. And we cook."

"Cook?"

"Yes. He enjoys cooking. He says it is gratifying to start with raw ingredients and have a tangible result in just an hour or two. He says he may never see results from what he does at work."

"How sad. What about you? Your work?"

"It's all rather boring, I'm afraid. The Air Force deactivated the flying wing last year."

"Do they need a weather officer if they don't have airplanes?"

"We still get transient aircraft, but not many. I think it is only a matter of time before they close the base."

"Will you have to move?"

"No, I will separate from the Air Force if that happens. My priority is Ernest. Oh, I hear his car. Got to go."

He walked in, tossed his hat and car keys onto a side table and trudged into the kitchen. "Smells mighty good."

"I hope it isn't burned."

"It will be fine," he said, reaching for her and drawing her into his arms. "It will fine from now on." He kissed her softly and looked into her eyes, which grew moist. "Something wrong?" he asked.

"No, not at all. I'm just surprised. You seem to be your old self suddenly."

"Old self?" He smiled and reached for a dinner plate. "I guess I am my old self."

They sat and he ate as she watched. "So, tell me!" she said.

"I've been working on a new design for our ECM, all very hush, hush."

"Beeps and squeaks?"

"Automatic beeps and squeaks. Come to find out, the work has already been done. In fact, it was done before Linebacker Two. To think all the lives that could have been saved."

"But why?"

"I don't know. I guess it doesn't matter anymore. That was a year ago. I was pretty upset about it."

"I remember."

"Really? Did I take it out on you?"

"No, Ernest. You've always kept these things to yourself."

"That's good. I guess that's good security wise. Not so good, marriage wise."

"Don't worry about that, Ernest. So, things are better now?"

"Oh no, they are much worse!" he said with an upward inflection of his voice and a smile. "Much worse!"

"You seem happy about it."

"I am. We've had a program for mounting cruise missiles to the B-52."

"Like a SAM, in reverse."

"Exactly. We use the bomber to get the missiles within a few hundred miles of their targets and the missiles do the rest of the job without the need to put a crew over enemy territory."

"But what's to stop the bad guys from shooting the missile down?"

"That's where my idea comes in. All I can say is that the missile can use its radar to figure out where it is, and it can fly so low that the enemy radar will never see it."

"So why do you need the bomber?"

"You don't!"

"I can imagine SAC isn't too happy with that idea."

"They hate it!"

"And?"

"And," Ernest continued, "I found someone at MIT who wants me to take my idea and run with it. The job pays double what an Air Force captain makes, and we don't have to move." His cheek muscles pushed his eyes upward as he smiled. "I want to do it. Is that okay with you?"

Marjorie sat on Ernest's lap and hugged him tightly. She kissed him and smiled, tears in her eyes. "It's nice to have you back."

"Does that mean yes?" he asked.

"Of course."

53. The Report

Omaha, Nebraska

"I can't let you submit this," Colonel Snedeker said. "I think your brother's ordeal has colored your judgment." Snedeker thumbed through the document. Stephen had submitted it a week before and Snedeker kept delaying the meeting.

"Everything in there is true," Stephen said.

"Yes, the chronology, the citations, all that. But that doesn't take away from the fact the Eighth Air Force report is also true."

"But it isn't true, sir."

"That's your opinion."

"I backed everything up with citations and I have copies of all source documents."

"Look, right here," Snedeker said, pointing to a page in Stephen's letter. "You say the Eighth Air Force position that the original plan worked is wrong. That's opinion."

"How can you say that, when the plan was thrown out completely on Day Four?"

"It wasn't thrown out, it was adjusted. The plan had room for improvement, so they improved it."

"I cannot believe this!" Stephen said.

"Look, Stephen, as your friend. The Eighth Air Force report is going to be adopted, no matter what. Submitting this letter accomplishes nothing."

"It puts on record an alternative view. Letting the Eighth Air Force report go unchallenged runs the risk of telling future commanders that the right way to run a campaign is by micromanaging it from headquarters and ignoring input from the field."

"Well, this is going to kill your career. I cannot let you submit this."

"I have the right to do so."

"Yes, you do. But not with the imprimatur of this office. You can submit it, but you must remove your title as a member of the CINCSAC Action Group."

"I'll do that, sir."

"Dismissed," Snedeker said. Stephen realized his relationship with his boss had just changed, making him angrier than he was before. Accordingly, he saluted and left. He considered lining out the title under his signature and handing it to the secretary to retype. Instead, he borrowed her bottle of liquid paper and dabbed the offending line, blowing on it gently to speed the drying. He then went to the clerical office and handed the report to the orderly at the front desk. "Two copies, please."

It was not yet 5 p.m. but he didn't have any other appointments. Susan would be home, and it seemed like a decision that required them both.

It was a cold day in May, even by Omaha standards. The drive to Papillon, Nebraska was on a long two-lane highway bordered by farmland. The small township of Papillon was quaint in its own way but had little to offer other than a small grocery store. "I really don't like it here," Stephen said to himself.

As he walked into the house and crossed the foyer into the kitchen, he spotted Susan stirring a pot of spaghetti sauce, which meant it was Wednesday. "How did it go?" she asked.

"Not well," he said. "I'm ready to pull the trigger."

"I'm ready to stand by you all the way, Stephen. We have savings. We can do it."

The next morning Stephen went through his normal "go to work" routine as if on autopilot. By the time he got to Building 500, he hardly remembered the steps leading to that point.

He got off the elevator on the top floor and chose the route that led to CINCSAC's office first. He opened the door and presented his report General Ham's secretary. "For the last time," he thought to himself as he left the office.

Snedeker was at CINCSAC's twice-weekly staff meeting and most of his officemates were probably downstairs for coffee and donuts. He had the office to himself and it seemed to him to be as good a time as any. He picked up the phone and dialed.

"Jack, it's Stephen," he said. "You guys still want me?" He listened and smiled. "Well please consider me a volunteer, available thirty days from today." He listened and smiled some more. "I can't tell you how happy this is going to make Susan. Me too. Thanks, see you in thirty days."

That bridge crossed, he picked up his hat and walked out, stopping at the office secretary's desk. "If Colonel Snedeker is looking for me, tell him I'll be at the base personnel shop. I should be back before noon."

There were as many people at Offutt Air Force Base working inside Building 500 as outside, but the headquarters staff was still a tenant on the base and their personnel needs were handled by the local wing. The drive to the personnel shop took Stephen around the base's single runway where he saw an EC-135 "Looking Glass" airplane landing. He knew there would be a general officer on board, ready to take command of the Air Force's nuclear forces should the world go mad and start throwing nukes over the north pole. It had been a wild career in uniform.

The act of resigning a presidential commission as an officer in the Air Force had become quite easy once the war had ended. There were more officers than available positions, and having a major resign made room for captains to be promoted. The clerk simply had to make a call to a base in Texas, where another clerk made sure Major Stephen Bellum had no outstanding training debts that had to be paid back with more time in uniform, and that his particular career field did not have a personnel shortage.

"AFSC?" the clerk asked.

"What?" Stephen asked.

"Air Force Specialty Code," she repeated.

"Oh. One, One, Bravo, Two Charlie," he answered.

She repeated the code over the phone. "Okay, bomber pilot, got it, thank you." She looked at Stephen. "They are going to fax the paperwork to your office. All you need is for your boss to sign it and it will be done, effective the last day of this month."

Now things really did feel final. He returned to his office, which was now fully manned, and the secretary handed him his faxed separation forms. "Really?" she asked. "Really," he said.

Colonel Snedeker heard his voice and called out, "Major Bellum!"

"Yes, sir," Bellum said, appearing at Snedeker's door.

"Enter and close the door." Stephen did so and Snedeker pointed to a chair opposite his desk.

"CINCSAC took your report and threw it in the trash," Snedeker said.

"I'm sorry to hear that, but I figured it might end up that way."

"So did I," Snedeker said. "You know we can't have you working up here anymore."

"Yes, sir."

"I found a vacancy for you at the base historian's office. You can bide your time there until you retire."

"No need for that, sir," Stephen said, handing over the separation papers.

"Wow," Snedeker said. "I didn't see this coming. You were promoted early to major. How many years?"

"Three years, sir."

"Well, I guess there is a first time for everything." Snedeker signed the form and handed it back. "You understand my hands are tied."

"Yes, sir."

"It's really a shame it turned out this way, Stephen. This will have zero impact on the command. Zero."

54. First Person Research

En Route to Washington, D.C.

"The United States is vast," Ha said from her window seat. "The maps do not do it justice."

"And we are only halfway there," Martin said.

"Everybody gets along, it seems. Since the civil war, that is."

"Pretty much. The states have their own power, the federal government has its power, and they both are subservient to the people. In theory, anyway."

"Democracy, truly."

"Sort of. It is a democratic republic. Democratic in that we elect people to represent us. Republic in the sense that the business of the government is the business of the public."

"Much better than communism. Or other -isms."

Martin let her statement go unchallenged. He knew her ideas were naïve. But he could no longer articulate the reasons they were wrong. She seemed relatively calm for someone with so much to lose. He realized that he might have been more nervous than she. It may have been only his second airline trip without his parents, and he had never been to Washington, D.C. It was a high stakes gamble for her. For him? Was it anything more than an academic exercise for a history professor? No, it was much more.

"Was Vietnam really communist?" he asked.

"They never spoke of it as such. It was the 'party' or the 'workers' they spoke of. I remember the pamphlets they had circulated after Tet in 1968. They promised that once the evil of the American puppet government was expelled, the common farmer and worker would have all power. But my sisters tell me we have replaced a corrupt government with a corrupt and vindictive government."

"Vindictive?"

"Yes. The new government seeks to punish all those who did not support them prior to the unification of north and south. That is why so many are in prison. That is why my parents are in prison."

"I've only read about such atrocities. It seems like something you read about in ancient times. It is," he paused, searching for the right words. "It is very sad."

"Yes, it is so."

55. New Hire

Denver, Colorado

"I talked to your old boss, Colonel Snedeker," the United Airlines training captain said from his side of the table. He was flanked by two other captains, a flight surgeon, and a human resources officer. The captains were in their airline pilot uniforms, the other two in coat and tie. Stephen was wearing the only suit he owned that still fit. His tie was cutting the circulation to his brain and he could feel his pulse in his neck.

"What can you tell me about him?" The captain asked. The five United officers each looked at Stephen. His mind raced. *Snedeker?* What does Snedeker have to do with him becoming a United Airlines pilot?

"I learned a lot from him," Stephen said. "He has a pressure cooker job and has to balance the push and pull from the SAC Operations Staff with the needs of CINCSAC. He does a great job of it."

"No hard feelings, having to leave?"

"No sir," Stephen said. "I know I could have kept at it for the remainder of my twenty years in the Air Force, but I was afraid my enthusiasm for the job would suffer and with that my performance. Whatever I do, I want to be one-hundred percent into it."

"Like flying airplanes?"

"Yes, exactly like that."

"Bomber pilot?"

"Yes, sir."

"Any good?"

"I think so. I got the job done and always brought the crew home safely. But to be fair, in wartime there are things that happen outside your control. Getting blown out of the sky isn't a reflection on the crew."

"On whom is it a reflection of then?"

"Fate."

The United officers exchanged glances. Stephen thought that they almost looked to be under as much pressure as he. It was a ridiculous thought, he concluded.

The lead captain returned his gaze to Stephen. "When was the last time you actually piloted an airplane?"

"Three years ago, when I left U-Tapao."

"How did our simulator feel?"

"Great! Even though the B-52 is a Boeing, the 727 feels much more responsive in all three axes. If the simulator is an accurate representation of the airplane, I am betting she is a dream to fly."

"She is that. How do you think you did?"

"I think my instrument procedures are rusty and I might have been a little rough on the controls. Nothing a little time in the seat can't fix."

"Well, here is what the flight examiner said after his ride with you." He fumbled with a manila folder and flipped a few of the pages. "He said you have the roughest hands he's seen in a long time and that you don't know the first thing about how to enter a holding pattern."

"Oh." Stephen's hopes sank but, as his father often told him, things happen for a reason and losing one battle was not the same as losing the war. The lead captain flipped a page in the report.

"But he also said you fly deliberately and that you have a good situational awareness of the airplane and its place in the air. I suppose flying in cell formation while being shot at does that to a pilot."

"Yes, sir."

"I asked him if he would hire you, if it was up to him. He said no."

"I am sorry to hear that."

"But it isn't up to him. I asked him how much combat time he has, and he said none. I asked him if he thought our veterans deserve a break now and then he said yes. I asked him if he would like to revise his answer, and this is what he said. He said that you are teachable when it comes to flying airplanes. He said it was up to me to decide if you are the kind of person we want flying our million dollar airplanes with hundreds of people putting their lives in your hands." The captain looked at the HR officer.

"The candidate seems remarkably intact, given what he's been through."

"I thought so too," the lead captain said. "That's why I called your old boss. He told me how you had the unmitigated nerve to challenge the entirety of the Strategic Air Command to do what was right. And that you lost that battle. He said he wasn't very nice to you on your way out the door and that you deserved to be sour about the whole thing."

"I am glad you told me that, sir."

"I'm glad you have conducted yourself so well. I told my colleagues here that if you started bellyaching about the way you were treated at SAC, we would tell you no thanks, say goodbye, and wish you well. But you didn't do that, proving what your old boss said. He said that men with character like yours are rare. And that when someone like you walks in the door, we should jump at the chance to hire you. Son, welcome to United Airlines."

56. Foggy Bottom

Washington, D.C.

"How is your room?" Martin asked.

"Very nice," Ha said. "I have only imagined such luxuries."

"It is the bare bones basic room here. I'm glad it is okay." He had dressed for dinner with a coat and tie, but instantly regretted that. He may have been the only person wearing a tie, other than the waiters. She was dressed as always, jeans and a nice floral blouse.

"How was your home in Saigon, before you left?"

"It was an apartment with two bedrooms. The three of us girls had our own room and our brother stayed with our parents. It was very nice."

"Has your family always been in Saigon?"

"Since I was very young. We lived in North Vietnam and became refugees after Dien Bien Phu."

"Refugee? When was that?"

"It was, I believe, 1954. The French government had left. My father says the Vietnamese Communist Party was killing all those villagers who did not show sufficient loyalty to the party. We escaped to the south, many kilometers."

"How awful," Martin said. He knew the basic events of the French Indochina wars, but it had never been of any scholarly interest. "You would have been just a baby at that point?"

"I think about four years of age. But I don't remember much of it. My first memories were in Saigon."

"Those were pleasant memories, I hope."

"Mostly. As my father had only a clerical job, he was not a target of the assassinations."

"Assassinations? Of whom? And by whom?"

"The South Vietnamese government back then was very corrupt. The Communist party killed over two thousand officials of the government. They started with the important people and gradually worked their way down to everybody."

"But not your father."

"No, by then, the Americans had arrived, and the communists left the city."

"I imagine you were happy to see the Americans."

"Yes, it is so. But the corruption in our government remained. The communists moved to the countryside where the killings continued. It is very sad." Her voice trailed off. Martin regretted bringing the subject up in the first place.

"Your father sounds like a very good man."

"It is so. Things were good until Tet."

"In 1968?"

"Yes. The war then came to the cities. It became normal to see corpses in the streets and then came the cholera. My father's duties expanded to giving the serum inoculations. It became apparent that life in Saigon would get worse, not better. That is when my parents managed to get the Student Visas for my brother and me."

"I hope we can return his favor to you by bringing him and your sisters to the U.S."

"Yes. I am so grateful."

"Let's not get too far ahead of ourselves. Tomorrow is a big day."

"Yes, it is so."

* * *

"Welcome to the State Department, Miss Tàu. I am Donald Wilson, one of Martin's old classmates. How was your trip?"

"Very nice," Ha said. "It was very comfortable, but very long."

"Yeah, coast-to-coast is definitely long," Donald said.

"I really appreciate this, Don," Martin said. "What are our odds?"

"They are better today than they were when we talked last week," Donald said, smiling. "Miss Tàu's neighbor in Manila has been quite helpful. He has what can only be described as a photographic memory. He recounted nearly fifty people whom we've now confirmed are interned at the worst re-education camps. Let's head to my office first, we have about an hour before the Ambassador arrives."

"Ambassador?"

"Well once you are an ambassador the title never leaves you, I guess. He is the Honorable Maynard Olsen. But you address him as Ambassador."

Ha and Martin nodded. "Good to know," Martin said.

Donald's office was a small five-by-ten affair with a desk and two extra chairs. With the three seated, their knees almost touched. His one window faced a parking lot shared with the post office, probably the only less than scenic view from the entire building. "Cozy?" he asked.

"It is very nice," Ha said.

"I wanted to spend some time here before we see Ambassador Olsen. I have some good news with some bad news, and a game plan I wanted to go over." Martin and Ha sat quietly. "Miss Tàu we have confirmed your father's location in the Le Van Duet prison in Saigon. Hanoi acknowledges he is there."

"It is so nice to hear that!" She said, tears streaming.

"Why is that good news?" Martin asked.

"Because if they acknowledged he is alive," Donald explained, "that means they cannot now make him disappear. That is the good news. But it is a small prison built for two hundred. They have in there right now over two thousand."

"Two thousand?" Martin asked. "The new government is imprisoning two thousand members of the former government?"

"Yes, more than that, actually," Donald said. "Le van Duet is just one prison of many. And that leads us to the bad news. The bad news is we have been unable to confirm the location of your mother, Ha. That isn't terrible news because they do not track the women prisoners as well as the men. It could very well be that they have her, but even they don't know where she is."

"I see." Ha appeared happy and relieved to hear of her father's fate but resigned to that of her mother's.

"You mentioned a game plan?" Martin asked.

"Yes. This is touchy because the government in Hanoi doesn't formally admit these re-education camps even exist as prisons. But we managed to get them to release two prisoners to us so far, based on medical needs of their relatives here in the U.S. It was a goodwill gesture on their part. The Ambassador thinks we can use this to expand the number of people they are willing to send us, using the idea of reuniting families."

"That is great," Martin said. "Ha and her brother are here; their sisters and parents are there."

"Not that easy," Donald interrupted. "The problem is that you, Miss Tàu, and your brother have no citizenship claim in the U.S. You are foreigners on Student Visas, that is all."

Martin slumped back in his chair; Ha looked to him.

"But you have a plan?" Martin asked.

"Well, an idea, the plan is up to Miss Tàu."

"Okay, let's have it."

"Time is against us, but I think this could work. If Miss Tàu were to get married to a U.S. citizen between now and when her Visa expires, and then be approved for the program the Ambassador wants to unveil this summer, then I think we can get her parents and maybe even her sisters to the U.S. There is a helluva lot of red tape to negotiate, but . . ."

"Married?" Ha said. "I am not married."

"He means you need to get married before you graduate," Martin said.

"Who will marry me?" she asked. "I do not know anyone willing to do that for me and my family."

Martin sat back in his chair again. The solution was obvious. But it was insane. "This has to be done right now?"

"I think it would be better," Donald said. "The Ambassador was excited to hear about this case. It is fresh in his memory. If he approves, we can get this plan written up and once it is in writing, it takes on a life of its own."

"You had this in mind all along, didn't you?" Martin asked, facing Donald.

"Kind of wild, isn't it?"

"Yeah. Could we have a moment?"

"You bet. I'll be back in thirty minutes."

Donald left the office and Martin turned to face Ha.

"You will do this for me, Professor Bellum?" she asked.

"Only if you call me Martin," he said.

"But why?"

"I am a flawed man who has been wrong about so many things. I want to make things right. I want to do something genuinely good for someone other than myself. I want to do that for you and your family."

"But marriage?"

"I know. I am willing to do this until the day you decide you can no longer stand me. But I hope that day never comes. In my family, the Bellum family, marriage is sacred. It is for life. I am willing to make that commitment."

She looked into his eyes for just a few seconds and then made her decision.

57. April 4, 1976

Lexington, Massachusetts

Finishing the tour, Ernest showed everyone into the living room where a fireplace stood ready for the senior Bellum's fire building skills. "You got the logs all wrong, Ernest. Let me fix this for you." Jonathan moved his prosthetic leg to one side while kneeling on the other leg.

Marjorie brought in an additional chair so everyone could be seated. Stephen took the chair, leaving space for Marjorie to join Susan and Mary on the couch.

"Okay," Jonathan said, "all set." Ernest handed him a long match and within seconds the fireplace was ablaze. He helped his dad up and into a recliner.

"It is a lovely home," Mary said. "All this on a captain's salary?"

"Two captains," Stephen said. "Marjorie is a captain now too."

"Oh congratulations, dear."

"Not for much longer," Marjorie said. "We have plans."

"Sounds ominous," Stephen said.

"Not really," Ernest said. "I am turning down my promotion to major and separating in a few months. I got hired by MIT as a research fellow."

"How's that pay?" Jonathan asked.

"Better than the Air Force," he answered. "In fact, if Marjorie wants to quit, she can."

"Will you?" Mary asked.

"Probably," Marjorie said. "The Air Force closed its flight operations here and I think the weather shop may be closed soon too. If they reassign me someplace else, I'll separate from the Air

Force. My priority is Ernest and hopefully we can start our own family soon."

"I am very happy to hear that," Mary said, looking to her husband.

"Me too," Jonathan said. "The two of you should settle down. I'm not saying having a husband and wife in uniform is wrong, but this particular husband and wife have served above and beyond the call of duty. You both deserve to be happy."

"Well as long as we are breaking news," Stephen said, "I have my own news. I just got hired by United Airlines. I start next month."

"Good for you," Ernest said. The family instinctively looked to Jonathan, who seemed lost in his own thoughts.

"Dad?" Stephen asked.

"Good for you, son," he said finally. "I thought the Air Force was right for you and I think you've done good things for the Air Force. What you did on the SAC Operations Staff took courage and saved lives. You did good. But you cannot oppose the will of SAC without collateral damage. Get out. Get out while you can."

Susan smiled and clenched Stephen's arm. "I told you!" she said.

"Now that we got that sorted," Ernest said, "we need to figure out what's going on in California."

* * *

The fireplace needed more attention and the three men set about restoring order to the flames. Marjorie got up and gestured to Susan and Mary, pointing the way to the dining room. "We knew it would end up this way, didn't we?"

Seated with their coffees, Mary finally said what she had been thinking since she arrived. "You two look so very happy, like the day we met you."

"We are," Marjorie said, beaming. "It was like a switch. Ernest came home from work and asked me if I wouldn't mind being married to a civilian research fellow. I told him it would be a dream come true. He put in his papers the next day."

"It seems like a huge weight has been lifted from his shoulders," Susan said. "Kind of like what happened to Stephen when he put in his papers."

The two younger women smiled at each other, realizing they were sharing in the same joy. But at the same time, it seemed Mary wasn't so sure. "Mary?" Marjorie asked. "Is anything wrong?"

"No, of course not," Mary answered. "They are both obviously happy with their decisions. I just hope they stay that way. Stephen and Ernest both become very attached to their causes; if you know what I mean. It is unlike either of them to give up on anything."

Marjorie and Susan began to speak at the same time, stopped, and laughed. "You first." "No you first."

"Okay," Marjorie said. "Ernest can't talk about it other than to say he has a plan that is going to force a solution onto SAC, whether they want it or not."

"Really!" Susan said. "Stephen says the exact same thing with his Linebacker Two history."

"Now I do feel better," Mary said. "I knew my boys could never give up a battle."

58. Disclosure

London, England

He felt almost like a cloak and dagger spy, walking through Mayfair, London up along Curzon Street, clutching a briefcase of stolen photocopied documents. The feeling was amplified as he walked across the street from Leconfield House, the home to the British Security Service / Military Intelligence, Section 5. But his destination was a café in Belgravia.

It was a corner establishment, with a dark wood exterior lined with leaded glass windows. The narrow sidewalk was made impassable by the tables with umbrellas and dining tourists. Entering the establishment, the dark theme was continued. The room was dominated by a bar shaped parallel to the exterior walls, giving way to a larger room with tables in booths. He spotted his contact, already seated at a booth at the far end of the bar.

He was as Stephen remembered him: in his late fifties, thick glasses accentuating a long, narrow face, and the physique of an athlete gone to pasture. He could have been mistaken for a government bureaucrat, atrophied by years of sitting at desks and working ledgers. But having talked with him for hours a year before, Stephen knew better.

"Major Bellum," the Brit said, rising and extending a hand.

"Mister Bellum now, but please call me Stephen. Good to see you again Michael."

"It is mutual, my friend. I took the liberty of ordering a bitter for myself. Can I offer you the same?"

"That would be great."

Michael left and returned, followed by a barmaid carrying two pints. "Cheers," Michael offered.

"Cheers."

Stephen considered the dark amber brew and the creamy white head. "This is good, but I thought you English drank this warm."

"American stereotypes," Michael said. "A bitter is nothing more than a pale ale served at cellar temperature."

"It is very good."

"So, you are no longer in the Air Force, I take it."

"United Airlines now. I am doing JFK to London these days. I bid this trip specifically to see you. Thank you for making the time."

"I was quite surprised to get your call, after the reception I got last year. Quite surprised."

"I apologize for that," Stephen said. "I know how you were treated, but the SAC Operations Staff has a particular view of Linebacker Two that is, how shall I put this? Do you know the phrase, 'history is written by the victors?'"

"I do indeed."

"Well SAC is writing history ignoring a mountain of evidence. I spent the last two years trying to fix all that, but I lost that battle."

"I've seen the official narrative that has been released recently. It is, shall we say, creative writing?"

"It's B.S., that's what it is."

"How do we correct that?"

"I come bringing presents," Stephen said, lifting his briefcase. "None of this is secret, it is all unclassified transmissions from the wings at U-Tapao and Andersen to the Seventeenth Air Division in U-Tapao, to the Eighth Air Force in Guam, to the SAC Operations Staff in Omaha."

Michael thumbed through the hundreds of pages and whistled. "They'll have your head for this if they find out."

"Nah, it's America," Stephen said. "But even if they do, it will be worth it. We cannot let them rewrite history. The Air Force has to learn the lessons of what happens when they let a major air command run a war without checks and balances."

"Is that what happened then?"

"Well, there was more to it than that. Pay special attention to the communications from the wings to the 17th Air Division. I didn't know this at the time, but almost every solution we ended up using after day four, was already figured out by someone at the wings. The bureaucracy was too scared to push the ideas to SAC. And when they did, SAC ignored them."

"According to the official report, the original SAC plan worked," Michael said. "The plan itself allowed for changes. But those changes took time to filter up to Omaha. But once they did, the changes were made. Linebacker Two ended the war for the United States."

"My brother has a theory about this."

"Your brother?"

"He is, I mean he was, an Electronic Warfare Officer in a B-52 that got shot down on Night Three. He was guest at the Hanoi Hilton for the rest of the war."

"I'm sorry to hear that. I trust he is okay now?"

"He is, in fact, he's getting out of the Air Force too."

"And he has a theory."

"Yes. Now Ernest is very smart. He's one of those guys who can solve a puzzle even before all the clues are out. He says Linebacker Two was like a very large math problem where we had more variables than equations. I'm not a math guy myself, but Ernest says you can't solve a math problem until you have at least as many equations as variables and this is a problem where we don't."

"You are losing me."

"Yeah, me too. Let's put it this way. Ernest says we can solve what happened during Linebacker Two, but we are still missing a few things. He thinks the fact we changed tactics after Night Four did shape the results. No doubt about it. But he also thinks that even had we done nothing, the results might have been the same because the North Vietnamese were running out of missiles."

Michael had started to lean forward as Stephen explained the theory, but when Stephen got to the punchline, Michael fell backward into his chair.

"Stunning," he finally said. "I just interviewed a Russian defector who was in Hanoi during Linebacker Two. His job was to train the North Vietnamese missile radar operators." He took a pull of his bitter and sat, not bothering to wipe the foam from his upper lip.

"And?"

"And," he said finally, "this defector told me precisely that."

Now it was Stephen's turn to sit back, stunned. Ernest was right again. "Michael, you don't have to tell me another word, except that you are going to publish this. But, like I said, my brother was not only an EW, but he was a POW. He came this close," Stephen said while pinching two fingers within a millimeter of each other to demonstrate, "to getting sent off to Russia. They would have sliced and diced him, and we would never have seen him again."

"No doubt," Michael said.

"So, you would be doing me a personal favor, and my brother, if you could fill in some of the gaps for us. Give Ernest some of his missing equations."

"Yes, indeed," Michael said, shuffling through the papers. "It seems more than fair. What do you want to know?"

"Well, first of all, how were they able to fire their missiles without firing up the Fan Song first?"

"I am not sure; it could be that you have some information here that will fill in the gaps that I have. But it appears they were tracking the routes using search radars and doing the maths to figure when they would be overhead Hanoi. They made it an A-level maths problem."

"Hard to see them getting that accurate enough to hit targets moving at five-hundred-plus miles per hour."

"You Americans do not have a monopoly on computers, but your point is well made. I think there is something to be said about the fact each airplane was followed by two others, and that each group of three were followed by another group. You Yanks made it easy on them."

"That we did," Stephen admitted. "What other bone-headed mistakes did we make?"

"The one that comes to mind for me was the four-hour delays between each bombing wave on those first three nights. That was critical."

"How so?"

"Each Surface to Air Missile site had a battery of six missiles. After they shot those six, they had to get resupplied from depots where the missiles had to be put together. Can you guess the turnaround time from the depot to having a missile ready to fire?"

"Four hours?"

"Just under."

"We did make it easy on them."

"And why did it take you nearly a week to start targeting them?"

"Well, they are mobile. They kept moving them around."

"No, the depots. If you took those out on the first nights, you would have cut the number of B-52 losses in half. At least."

"Yeah," Stephen said. "We kind of figured that out the hard way."

"From our talks last year, it seems you already figured out the problem with the pre and post target turns."

"Sort of. We figured out the post target turn alright. It sends the ECM for that bird off to the horizon and reduces cell jamming. But nobody considered the pre-target turn problem until Ernest brought it up when he got back."

"Your brother came up with that?"

"Yes, the way he tells it, he drew it on the ground when he was a POW. Says he knew the third airplane was in the worst position because it ended up with the least amount of jamming when the first two airplanes turned away. But that didn't explain the SAM hits before the post target turn."

"It takes a while for the airplane on the left to catch up with the others following a right turn, doesn't it?"

"It does indeed. Sometimes the most complicated things seem simple, after you think about it."

"War is both complicated and simple. We figured that out the last time we went to war with you."

"Never again, my friend," Stephen said.

"Never again, mate."

59. April 4, 1977

Bellevue, Nebraska

It was her, just a little late, but she came. Stephen rose and noticed Jonathan struggle with his cane; he stepped over and helped steady his dad as he stood. Both rose to give Marjorie a polite hug.

"Happy anniversary, Marjorie," Jonathan said.

"Thanks dad, and happy repatriation day to us all."

"So happy you could make it back to Nebraska. I'm not traveling so well these days. Where is Ernest?"

"Ernest saw someone in the neighborhood he needed to talk to. He'll be right in."

"I never knew Ernest had friends in the neighborhood," Jonathan said.

"I thought Marjorie was the only friend he ever had," Stephen said.

"You two!" she said. "Ernest has a lot of friends, thank you very much! Please sit, dad. I hear the prosthetic leg has been bothering you lately."

"Nah," he said. "I'll be fine." But he sat down all the same. Marjorie and Stephen joined him.

"So," Stephen said, "this makes three years of marriage and four years for Ernest with his life back."

"And almost a year of civilian life for the three of us," she said. "How is the glamorous life of an airline pilot?"

"It's glamorous."

"You don't miss the Air Force? At all?"

"Maybe a little," he admitted. "I miss the sense of having a purpose in life greater than me. I guess my purpose now is my family."

"You deserve that. I feel more like a part of the family than the one I grew up with, thanks to all of you. Well, all of you minus one. I wish Martin would join us one of these years."

"Don't we all," Jonathan said. "Last I heard from him was after he graduated from college. Nothing since."

"Everything happens for a reason," Stephen said. "You taught all of us that."

There was a commotion at the front door and a joyous scream. Ernest came to the back porch. "There is someone at the front door you should see. It's Martin."

Martin stood in the entryway, unable to release himself from Mary's hug. Susan and Marjorie stood close by, both with tears in their eyes. Martin gave his father and brothers a smile, knowing they would understand his helplessness. He was dressed somewhat formally, dress slacks and a tweed jacket. He had put on a few pounds, almost hiding the young Vietnamese woman standing to his side. She was considerably shorter, wearing a colorful sundress and clutching a wrapped present. She smiled and stood to one side as Jonathan approached.

Martin smiled, but seemed unsure of what to do next. His eyes moved down to his father's wooden leg and back to his eyes. He extended a hand, which his father took to leverage the rest of his son in a hug. "Son."

"Dad."

"Family," Martin finally spoke, "this is my wife, Ha. Ha, this is my, uh, our family." The Bellum women hugged some more and all four spoke at once as the men stood, saying nothing.

"Let's sit out back, boys," Jonathan said, leading the way.

There were three padded chairs and a padded rocking chair, which was reserved for Jonathan. He pointed to his outdoor refrigerator. "Stephen, it's five o'clock somewhere in the world." Stephen went to the fridge and pulled four beers and handed each out. He sat down and spoke first.

"Who would have thought that between the three of us brothers, you would be the first to go gray."

Martin blushed. Stephen's hair was no longer jet black but had settled on a dark brown. Ernest's hair looked as it did when he was a teenager, maybe a little thinner. Martin's hair was mostly dark, but the speckled gray took over along the sides down to the long sideburns.

"I dye it," he finally admitted. "It is hard to be a college professor when you look younger than your students. Tricks of the trade, I guess."

"Is it all you hoped it would be, son?"

"The professor part, yes. But I think I've learned more about history and myself in the last year, than all the years before it." Before he could say anything more, Susan entered, carrying a chair from the dining room. She was followed by the other three Bellum women.

Mary hugged Martin again, reaching from behind his chair. "Jennifer, Martin that's Stephen's oldest, Jennifer volunteered to watch the boys so we could all sit and talk."

Martin smiled, looked at Ha, and took a sip of his beer. Ha looked at him and put her hand on his. "Martin, it is time," she said. He nodded. She sat with her legs drawn together, her hands neatly folded on her knees, and her eyes focused on Martin.

"I have a speech I have been rehearsing," Martin said. "We drove all the way from California and Ha has listened to me give it over and over again. I talk all day for a living. I give lectures in auditoriums with hundreds of students. You would think this would be easy for me." He unfolded his hands and placed one on each knee, as if to steady himself. "When we finally got here, I kind of froze and circled the block. I parked down the street, trying to get the courage to walk in the front door. And then Ernest drove up, spotted me, and came over."

"You can't imagine how happy I was to see you," Ernest said.

"Well, having you here makes my speech all the more necessary," Martin said. "So here goes. When the people from the Air Force came into my dormitory to tell me Ernest had been killed, all hell broke loose. Things are different these days. But back then, the campus wasn't the friendliest place for anyone in uniform. When they came to see me, telling me the Chief of Staff of the United States Air Force extended his deepest sympathy, well, I snapped. I threw them out."

Ha held him by his trembling arm. "Of course, I later realized it wasn't those people I was mad at, it was the war. Even though I was a history major who thought I knew all there was to know about war, I didn't. But as far as I was concerned, everyone and everything connected to the Vietnam War was evil."

"I filled out the paperwork to become a conscientious objector and I am ashamed that I timed mailing it waiting for dad's check to clear on my last semester of undergraduate college. When the news came out that Ernest was alive, I was relieved. Of course, I was relieved. But it made me even madder that they told us he was dead. And that," he paused to choke back a sob. Ha squeezed his arm.

"Well," he continued, "it only made me madder. I vowed to never have anything to do with the military again and when they cancelled the draft call up, that became easier. But you, my family, became collateral damage. As the military likes to call it."

He looked at Ernest. "Of course, when you came home, I had to see you. Had to. But that was the last exception I made. It became my life's purpose to teach the world the evil of war. No matter the reason."

"I was doing exactly that, two years ago when I met Ha. She was on a Student Visa at Berkeley with her brother. The family plan was for them to get established and find a way to bring their mother, the father, and two younger sisters here. All that changed when Saigon fell."

"What happened to Ha's parents?" Mary asked.

"Well that is the point of all this, I guess," he continued. "I was teaching that no matter who started it, because of the repercussions, any participation in war is wrong. I had excuses for everything. If we never joined with the allies against Nazi Germany, they would have limited their aggression to Europe, the

Europeans would eventually rise up, and it would have been over with far quicker and with fewer lives lost."

"But what about . . ." Jonathan started.

"I know, I know. I know that now. But this was two years ago. And that's where Ha comes in. After I got drafted, I started to think that maybe we should leave the Vietnam problem to the Vietnamese. It was their country, and it was none of our business. The Russians would never have gotten involved if we had never gotten involved. No SAMs, no B-52s, Ernest never gets shot down. It made sense to me at the time. But then I met Ha. Her father was a town clerk for the Saigon government. But he worked for the former government and that wasn't good. In fact, it became a crime. He and Ha's mother were put into re-education camps the same year Ernest was released from the POW camp. It's like they exchanged one set of prisoners for another."

Ha brushed a tear away.

"One of Ha's neighbors escaped and brought with him news that Ha's father is alive. He also spoke of the living conditions. They are not good. They are torturing some of their prisoners. We thought there were 300,000 in these re-education camps. Now we hear it is closer to 500,000. We got to talk to him, ah, the neighbor who escaped in January. He says at first the reeducation camps were mostly classrooms and tours of B-52 crash sites. Who knows, maybe one of them was Ernest's airplane. But after a few weeks, the prisoners were sent off to do hard manual labor around the towns. Some were put into solitary confinement and never seen again. He escaped while on a work detail. Talk about hell, he's been through hell. He says his parents were made to testify against each other by withholding food until they agreed to say the other was an enemy of the party. As a reward they were put in less harsh conditions, but they are still not permitted to leave the camps."

Marjorie reached for a box of tissues which was quickly passed around.

"This won't take much longer," Martin said, "I promise you that."

"Take all the time you need, son," Jonathan said. Martin offered a weak smile, but he continued.

"I met Ha and she opened my eyes to what it was like in South Vietnam while we, the U.S. were there and how it has become ever since. All these years I was harboring all this venom towards the United States government and the military. And that included you, dad, and Stephen, and Ernest. Then things changed two years ago."

"Ha was facing deportation as soon as she graduated, when her Student Visa expired. I couldn't believe it, so I called a friend of mine at the U.S. State Department. He and a deputy director came up with a new program to convince the Vietnamese government in Hanoi to release some of the prisoners and allow them to bring their families to the U.S., provided they had a family here or a relative married to a U.S. citizen. The State Department came up with the program with us as the test case. They did that for us."

"That's wonderful," Mary said, wiping a tear.

"By then I already had a good idea that the war wasn't a war against the people of Vietnam," Martin said, "but it was a war for something that might have been too intangible to understand. The same government I had been cursing went to bat for us. And, just a few months ago, the Vietnamese government declared Ha's father too old to be a threat to the party. They agreed to release her father to the U.S."

"Yes!" several of the Bellums erupted. Mary let out a sob, Marjorie and Susan held her tight. Martin continued. "That's when I knew that I have been an ass all these years."

"No," someone said.

"Yes," he continued. "We don't have re-education camps here and the way they treated Ha's family is criminal. We are not the criminals. Ha thinks Vietnam will figure it out eventually. But we'll get her dad and sisters out so they can live in the freedom of the United States."

"What about Ha's mother?" someone asked.

"She passed away last year," Ha said. "We do not know how. I imagine her heart was broken. My sisters were made to denounce each other as enemies of the party and then placed into different re-education camps. But the State Department thinks they will be freed as well."

"I'm so sorry for all that has happened to you and your family, Ha," Jonathan said. "But I am so grateful that you and Martin found each other and that you have joined our family. You are one of us now."

"Yes!" the group said collectively.

"Thank you," Ha said. "Enough with the tears, let us celebrate the day. It is Ernest and Marjorie's wedding anniversary, after all."

* * *

The gathering broke into two groups, those who favored the kitchen and those who favored the back porch and Jonathan's beer refrigerator. Marjorie started with the women in the kitchen, as seemed to be expected of her at most social gatherings.

"Martin says Ernest is the smartest man," Ha said. "The smartest of all men."

"I hear that a lot," Marjorie said. "He is pretty smart, but you have to peel back his outer layer to appreciate the man inside."

"He has never been easy to talk to," Mary said. "Even as a boy to his mother. He refused to give me straight answers. It was a game to him. But you've changed him, Marjorie."

"No, it wasn't me," Marjorie said. "I noticed he did that to me until he got to know me. And he did it with everyone else, I thought, because he didn't know them."

"You thought?" Mary said. "As in used to think that but not now?"

"This was before he was shot down," Marjorie said. "He didn't talk much about his days as a POW until after he left the Air Force.

Then, all of a sudden, it all came out. He says that he knew that if his captors ever found out he was an EW, his name would never be published, and he would be sent to the Soviet Union. So, he needed to put on an act to convince them he was a navigator. But he couldn't pull it off without giving them something else to focus on. He was taught in training to offer passive resistance that wouldn't get him hurt but convince them that he wasn't worth spending any time on. So, he said he played, 'captain annoying.' It worked so well he decided to retire the role."

Mary laughed. "He spent a lifetime annoying me so he would be so good at it as a POW they would fall for it. Everything does happen for a reason! I will never again get mad when Jonathan says that to me."

"Ernest has a corollary to that rule," Marjorie said.

"Of course, he does!" Mary said.

"Ernest says the reason isn't necessarily a good one. But good or bad, the reason exists."

"Martin says the same thing," Ha said. "I think maybe the brothers have more in common than we know."

"It will be good for the brothers to spend more time together," Susan said. "That is if you can get more than a few words out of them."

"Like father, like sons," Mary said.

"I think so," Ha said. "So, Ernest is no longer flying these airplanes?"

"Oh, no," Marjorie said. "He's back at the college where he got his engineering degree. Maybe he's following Martin's path."

"That is wonderful," Ha said. "And how about you, Susan?"

"Stephen left the Air Force. It was a little bittersweet."

"Bitter?" Ha asked.

"Sorry. Bittersweet. Both good and bad. He was honored to serve his country during the war and feels that he had a positive impact in helping to end it. But it was time to move on. The airlines started hiring, so he made the jump. We are very happy now." Ha smiled at that.

"Is Stephen really happy now?" Mary asked.

"He is," Susan said. "I know we were both angry, but time heals all wounds. Every time he talked about his time at headquarters, he realized more and more that he may have had a role in changing the direction SAC was heading. He still regrets not acting quickly enough to prevent Ernest from nearly dying. But he now feels certain his actions may have saved lives."

"I think having Martin back will also help," Mary said.

"It already has," Susan agreed.

* * *

Later that night, when all conversation was exhausted, Jonathan broke out a few bottles of wine and spread the glasses among the adults. Once each glass was filled, he began.

"This isn't the toast, so if you feel the need to sample your glasses, feel free," he began. "I joined the Army Air Forces in January 1942, but they didn't have enough airplanes, so I got to be a college student for another year. By the time I finished training, the war in Europe was almost over. I flew my missions and dropped some bombs, but I never saw any flak or enemy fighters. After the war I felt like a fraud. There I was the bomber pilot vet, having never really been in any danger. But you can't pick your birthdate and I did my part. As you know . . ."

"Everything happens for a reason," the Brothers Bellum said in perfect harmony.

"Yes," Jonathan said. "So, when Stephen got sent off to war and did see flak and did see enemy fighters, well I really started to worry. But nothing could get to the B-52, so I started to worry less. Then Ernest went off to war and I repeated the process until November 22, 1972 when the SA-2 finally got a B-52. When Ernest got . . . well, son, I know it was hell for you. But it was hell for your mother and me too."

"All of us," a few in the group said.

"All of us," he agreed. "I learned that just because you believe something is so, doesn't mean it is. And even if it is, it could change. SAC had its purpose in the fifties. But when war changed, SAC refused to."

"I am guilty of that too," Stephen said. "I flew the war before it got deadly for bombers and based on that was blinded by success. I even had a brother in the heat of battle, and still I was blind until it was too late."

"Not too late," Ernest said. "I think things would have been worse without your actions."

"Too late for you, I'm sorry to say," Stephen said.

"I don't know about that," Ernest said. "But if we are all in the mea culpa mood, I can add to that. I was convinced the technology was invincible before the war and after it failed, I spent a few years after the war trying to reinvent the same technology. Now I'm working on something completely different."

"Which is?"

"I can't tell you. But imagine a world where we never have to fly over enemy territory until after the threat is gone?"

"How?"

"Well, I can only say that you don't need pilots in airplanes to fly them."

They all sat in wonder. Ernest was being cryptic again, but the clues were there. Jonathan finally spoke. "Well, this is one pilot who is all for a world where we don't have to fly where the bullets are! I don't know if all wars are worth fighting or if any of them are. I do know that men and women in uniform will have no choice but to fight the wars the cowards in government will get us into. If we can do all that without putting our crews in harm's way, I say that's a good thing."

"Hear! Hear!" the former crewmembers said.

"But now we are all here, together. Our battles have been fought. Ha, I wish your father, your mother, your brother and sisters could be here too, but since they're not, you will have to represent them in spirit. So, my toast is a simple one." He raised his glass, and everyone followed.

"The gift of life must be appreciated to be lived to its fullest. Here is to life."

"To life," they all agreed. Ha and Martin smiled at each other and held their gaze for so long the group noticed.

"What?" someone asked.

Martin "broke lock" from his wife's eyes and faced his family. "It is the translation of 'Ha' to English." He beamed, returning his gaze to Ha.

"It means," Ha said, "the kiss of life."

Author's Note

They say, "fact is stranger than fiction," but in this case, it would be apt to say, "fact is even more tragic than fiction." I've tried very hard to keep the chronology of what happened and when it happened as true to actual life as possible. The seemingly suicidal rules of engagement were really what constrained our bomber crews. Most sadly, the bomber losses are also factual. I agree that Linebacker II hastened the conclusion of the American involvement in the Vietnam War. I disagree that the Strategic Air Command deserves credit for the eventual success of Linebacker II.

I encourage you to dive into the further reading I have listed in the following pages. But before that, I've listed an actual chronology to whet your appetite to learn more.

Actual Chronology

June 18, 1965 – First B-52 combat mission sent from Andersen to Vietnam resulted in a midair collision between two B-52Fs, killing eight of the twelve crewmembers.

April 10, 1972 – First B-52 strikes deep into North Vietnam.

November 22, 1972 – A B-52 is shot down over southern North Vietnam, the first B-52 combat loss of the Vietnam War.

December 16, 1972 – The Paris Peace Talks formally break down with no new date for resumption.

December 18, 1972 – Night One of Linebacker II – 129 B-52s strike targets around Hanoi, in three waves, four hours apart. Three B-52s are shot down, two are heavily damaged.

December 19, 1972 – Night Two of Linebacker II – 93 B-52s attack the Hanoi area with same targets, times, and routes as the previous night. Several B-52s are damaged but none are lost.

December 20, 1972 – Night Three of Linebacker II – 99 B-52s are sent to Hanoi with the same targets, times, and routes as the previous two nights. Six B-52s are shot down during the first and third waves.

December 21, 1972 – Night Four of Linebacker II – 30 B-52s are sent, only D-models from U-Tapao, and those are sent with the same timing of the third wave of the previous nights. Three B-52s are shot down.

December 22, 1972 – Night Five of Linebacker II – B-52s strike Haiphong, away from Hanoi. No B-52 losses.

December 23, 1972 – Night Six of Linebacker II – B-52s strike far away from Hanoi. No B-52 losses.

December 24 – 25, 1972 – 36-hour standdown, SAC turns over mission planning to the 8th Air Force Headquarters, tactics revised.

December 26, 1972 – Night Eight of Linebacker II – 120 B-52s strike Hanoi and Haiphong from seven separate streams with multiple entry and exits routes and altitudes, in a single wave. One B-52 is shot down, another is damaged and crashes at U-Tapao.

December 27, 1972 – Night Nine of Linebacker II – 60 B-52s repeat the pattern from December 26. One B-52 is lost, another heavily damaged. Hanoi notifies Washington it will return to Peace Talks without preconditions.

December 28, 1972 – Night Ten of Linebacker II – 60 B-52s strike five targets in six waves. No B-52 losses.

December 29, 1972 – Night Eleven of Linebacker II – 60 B-52s strike. No B-52 losses.

December 30, 1972 – Bombing suspended north of the 20th parallel.

January 27, 1973 – The Paris Peace Accords are signed. The U.S. Selective Service announced there would be no further draft calls.

February 12 – April 4, 1973 – A total of 591 American prisoners of war are released and flown back to the United States.

April 4, 1975 – A Lockheed C-5 Galaxy crashes while attempting an emergency landing after an explosive decompression. It had departed Tan Son Nhut Air Base, South Vietnam destined for Clark Air Base, Philippines. On board were American caregivers paired with South Vietnamese orphans. Of 314 on board, only 176 survived. Among the casualties were 78 children.

May 1975 – Specific groups of Vietnamese are ordered to register with the new government and in June were ordered to report for "re-education." There were between 300,000 and 500,000 interred at these "re-education camps," some were tortured, some were kept in solitary confinement, most were required to do hard labor. Incarceration periods ranged from weeks to 18 years.

April 29 – 30, 1975 – Saigon Evacuated.

June 2, 1992 – SAC disestablished.

Further Reading

Boyne, Walter J., "Boeing B-52: A Documentary History," Schiffer Publishing Ltd., Atglen, Pennsylvania, 1994.

Davies, Steve, "Boeing B-52 Stratofortress Owners Workshop Manual," Haynes Publishing, Newbury Park, California, 2013.

Drendel, Lou, "B-52 Stratofortress Illustrated," Aviation Art, Inc., Venice, Florida, 2018.

Eschmann, Karl J., "Linebacker: The Untold Stories of the Air Raids Over North Vietnam," Ivy Books, New York, 1989

Michel, Marshall L. III, "Operation Linebacker II 1972: The B-52s are sent to Hanoi," Osprey Publishing, Oxford, UK, 2018.

Michel, Marshall L. III, "The 11 Days of Christmas: America's Last Vietnam Battle," Encounter Books, New York, 2002.

Nalty, Bernard C., "Tactics and Techniques of Electronic Warfare: Electronic Countermeasures in the Air War against north Vietnam, 1965 – 1973," Defense Lion Publications, Newtown, Connecticut, 2013.

Rochester, Stuart L. and Kiley, Frederick, "Honor Bound: American Prisoners of War in Southeast Asia, 1961 – 1973," Historical Office, Department of Defense, Washington, D.C., 1998.

Toai, Doan van, "Vietnamese Gulag: A Revolution Betrayed – The Shocking First Person Chronicle of a Former Supporter of the Socialist Republic of Vietnam Who Survived His Country's Descent into Totalitarianism," Simon & Shuster, 1986.

Website Sources

"AN/ALQ to AN/ALT Equipment Listing" http://www.designation-systems.net/usmilav/jetds/an-alq2aly.html

"Linebacker II 12/72" http://www.linebacker2.com/Page_2_59BI.html

Vietnamese Reeducation Camps – The Vietnam War "https://thevietnamwar.info/vietnamese-re-education-camps/

About the Author

James Albright is an average pilot with average stick and rudder skills, but has an above average desire to learn and instruct. He spent twenty years in the United States Air Force as an aircraft commander, instructor pilot, evaluator pilot, and squadron commander. After retiring as a lieutenant colonel, he went on to fly for several private and commercial operators as an international captain, check airman, and chief pilot. His logbook includes the T-37B, T-38A, KC-135A, Boeing 707, Boeing 747, Challenger 604, and the Gulfstream III, IV, V, 450, and VII.

His website, www.code7700.com attracts nearly five million hits each month and his articles have appeared in several magazines, most notably Business & Commercial Aviation.

Made in the USA
Monee, IL
12 June 2025

19288019R00154